BASIC BUILDING SERIES

Plumbing: *F. Hall*

Painting and Decorating

L. F. J. TUBB, C.G.I.A., F.R.S.A., F.I.B.D.

Formerly Senior Lecturer in Painting and Decorating
Guildford County Technical College

Revised Second Edition in SI Units

MACMILLAN

First edition 1961 Reprinted 1963
Second Edition 1967 Reprinted with corrections 1969
Reprinted 1972 with SI notations
Second Edition 1974, revised in SI units

Published by
The Macmillan Press Ltd
London and Basingstoke
Associated companies in New York Dublin Melbourne
Johannesburg and Madras

SBN **333** 15765 6

PRINTED BY Unwin Brothers Limited
THE GRESHAM PRESS OLD WOKING SURREY ENGLAND

Produced by 'Uneoprint'

A member of the Staples Printing Group

Contents

Preface to Second Edition (SI Units)

In this second SI edition of a book that has received a friendly reception among Technician, Craft and General Course students and their teachers, the opportunity has been taken to make minor revisions to the text, in addition to the main changes from Imperial to SI units of measurement.

The system of presenting, so far as possible, text facing relevant drawings, has evidently made for easy study; and although the information is necessarily very condensed, it is hoped that no important point has been overlooked. Comments and criticism with a view to further editions will be welcomed by author and publisher.

With the adoption of SI units by the construction industry, it was felt desirable to revise the second edition of this book in SI units. It is appreciated that the reader may still encounter measurements in the Imperial system for some time to come, so a simple conversion chart is printed below. Where conversion factors relate specifically to the text they are reproduced as a footnote on the appropriate pages.

Paint is now packed in metric containers, and these are listed below, together with the nearest Imperial equivalent

100 ml (1/4 pint)	250 ml (1/2 pint)
500 ml (1 pint)	1 litre (1 quart)

These are all about 12 1/2% smaller than the equivalent Imperial size. The following are about 10% larger

2·5 litres (1/2 gal)	5 litres (1 gal)
25 litres (5 gal)	

CONVERSION TABLE

		SI	*Imperial*		
0·04 in.	←	1 mm	1 inch	→	25·4 mm
1·09 yd	←	1 m	1 yd	→	0. 91 m
0·62 mile	←	1 km	1 mile	→	1·6 km
0·035 oz	←	1 g	1 oz	→	28·3 g
2·2 lb	←	1 kg	1 lb	→	453·4 g
			1 stone	→	6·35 kg
			1 cwt	→	50·8 kg
1·75 pints	←	1 litre	1 pint	→	0·57 litre
0·15 lb/in^2	←	1 kN/m^2	1 lb/in^2	→	6·89 kN/m^2

The Reasons for Painting and Decorating

PRESERVATION, SANITATION, AND DECORATION

If the materials used for building are left unprotected from the effects of hard wear and weather they deteriorate and decay. The most economical method of protection is by painting, as, compared with other types, it is cheap, can be done at any time without undue inconvenience to the usage of the building, and easily conforms to any shape of surface. Surfaces are also painted to make them clean and more hygienic and to enable them to be cleaned more easily, a fact which finds expression in the Factories Act 1938, which enforces the inside treatments of factories within specified periods, according to the materials used. Lime washing should be done every 14 months, water paints and distempers renewed every 3 years, plastic emulsions (P.V.A. type) every 5 years, and oil paints every 5-7 years. Buildings and articles which are well painted and colourful give more pleasure, enabling them to be enjoyed more fully, and are also easier to sell.

Paints are fluid mixtures which, when spread in thin films or coatings over a surface, dry hard and form tough but elastic skins. For inside work, the materials are made to dry harder than for outside work, which needs to be more flexible; but many ready-made paints are dual purpose in that they can be used both for inside and outside work. In any case, the directions on the tin should be carefully followed. Paint manufacturers make materials for almost every type of surface met with in building—wood, building board, plaster, concrete, metal, asbestos, stone and brickwork; and complete protection is obtained in the form of a solid barrier built up with different types of paint—primers, undercoatings and finishes.

PRIMERS are the first coats of paint applied to a surface. They are the most important, as they form the foundation for all the paint applied over them. They are made with materials which give them special qualities and it is necessary to select them carefully according to the types of surface to be painted; thus, alkali bearing surfaces such as plaster or brickwork will need a primer quite different from that for iron and steel, where the chief protection needed is against rusting or corrosion. But it is important that the surface should be perfectly clean and properly prepared before the priming is applied. UNDERCOATS are used to reinforce and build up the thickness of the film and obliterate the surface, the number of applications depending on the cost and type of work. On new outside work two undercoats should be given but with new work inside one undercoat is usual. FINISHES give the colour and degree of sheen required and range from decorative materials drying with a perfectly flat or lustreless surface to full gloss paints and enamels, which are useful both for decorative and protective purposes. Special products are also made to combat dampness in buildings, and marine and industrial atmospheres; these include chlorinated rubber paints, synthetic resin finishes, and bitumen.

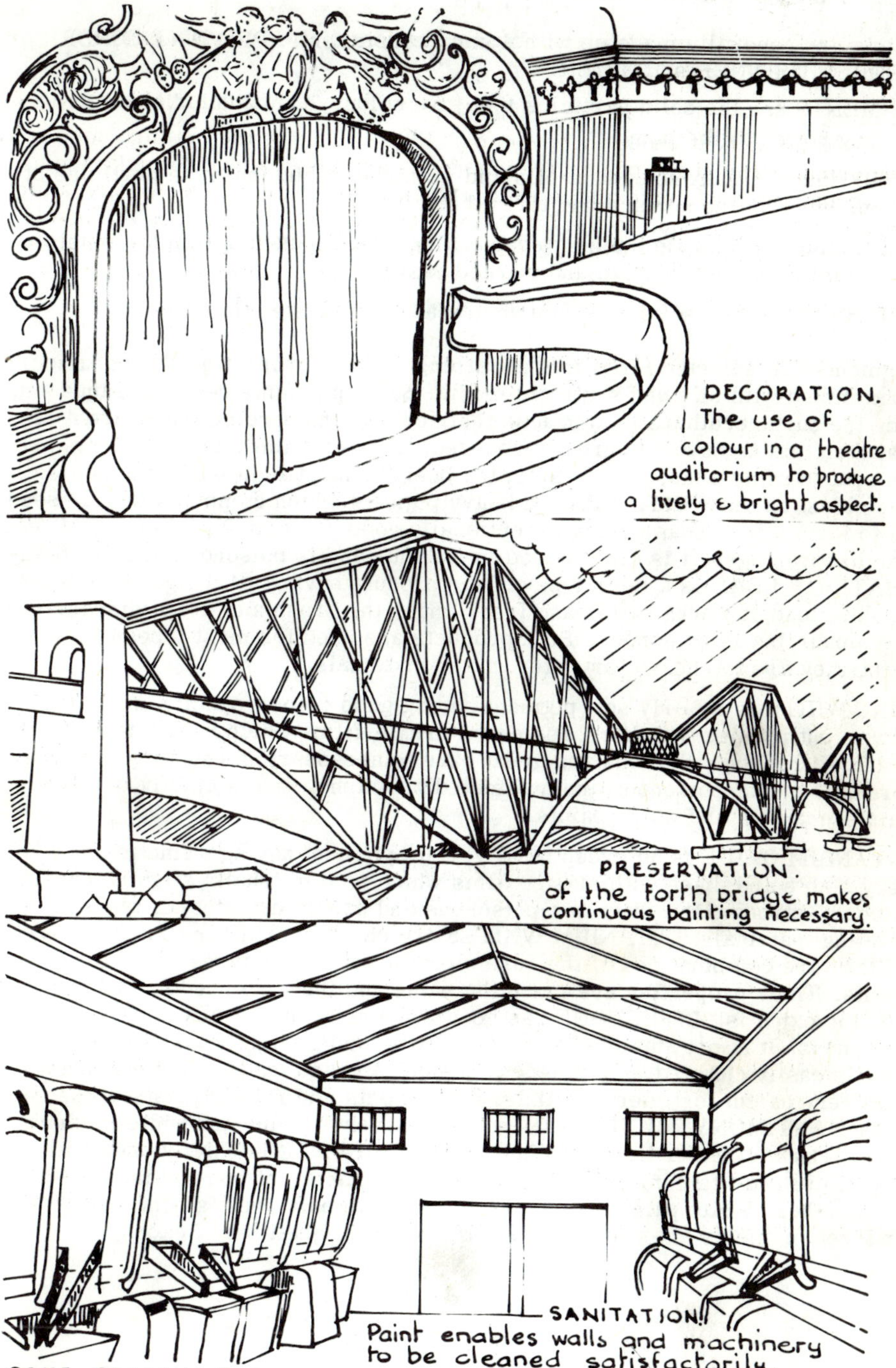

SOME REASONS FOR PAINTING - PRESERVATION. SANITATION. DECORATION.

Paints

THE CHIEF WHITE PIGMENTS

Paints are generally made up with 4 main constituents: PIGMENT, BINDER or MEDIUM, THINNER and DRIER.

PIGMENTS are the solid particles that give colour, body and hiding or obliterating power to the paint.

BINDERS or MEDIA consist of oils and varnishes which hold the particles of pigment on the surface and form the paint film.

THINNERS are volatile (quickly evaporating) spirits used to thin the paint, enabling it to be spread and worked easily.

DRIERS are either solid or liquid agents added to hasten the drying.

Pigments are selected for paints according to their properties and behaviour when mixed with oils and synthetic media. Some pigments remain opaque and hide the surface; others tend to lose their colour and become transparent so that their hiding power is greatly reduced. The most important pigments are the opaque white ones, as they form the basis of most light coloured paints. The oldest one is WHITE LEAD, a heavy pigment which combines with linseed oil to form a lead soap having exceptionally good protective properties. It has two disadvantages; it is discoloured by sulphur and is poisonous. BASIC SULPHATE OF LEAD is a later type of lead product having all the good qualities of white lead; it works better and freer under the brush and is not discoloured by sulphur, but is poisonous. Both products are used for most types of oil paint; they are too heavy for water mixed materials.

ZINC WHITE is a fairly soft pigment having good colour, opacity and covering power, but it dries with hard, somewhat brittle, films which have a tendency to crack. It reacts with acids and great care is necessary in choosing a media for it. It is non-poisonous and unaffected by sulphur and is chiefly used for paints, enamels and water paints.

TITANIUM OXIDE is an exceptionally white pigment having a fine texture. It is very stable and dries with elastic films which tend to 'chalk' slightly, causing fading of strong tints. It is non-poisonous and can be used for all types of painting materials. TITANIUM WHITES are cheaper than pure Titanium Oxide, as they are combinations of the pure pigment with other types of white pigments. Their properties are much the same and they are used for all purposes that the oxide is. LITHOPONE has good whiteness, and opacity, and is non-poisonous. It is not suitable for outside work, being slightly alkaline in reaction, is easily changed into soluble zinc sulphate, and tends to get washed away. It is very useful for interior paints, water paints and rubber paints, and as a filler. ANTIMONY WHITE is non-poisonous, very opaque and white in character, works well under the brush but is a poor drier. It is chiefly used with zinc white and titanium oxide to correct the hardness of one and the chalking of the other. White paints are seldom made with only one pigment; usually two or three pigments are used to obtain the inherent qualities of each.

PIGMENT.	DEGREE OF OPACITY	TYPE OF FILM.	DURABILITY.	DRYING PROPERTIES IN LINSEED OIL
White lead. Hydro-carbonate of lead.	4.	Tough and flexible.	Fairly good.	Quick.
Basic sulphate of lead.	3.	Flexible.	Good.	Average.
Zinc Oxide.	3	Hard and brittle.	Good.	Very slow.
Antimony white.	2	soft.	Fairly good.	Bad drier.
Titanium dioxide.	1	Tough and flexible.	.Good.	Average.
Titanium white. Mixture of oxide and other white pigments	2.	Tough and flexible.	Good.	Average.
Lithopone.	1	Hard.	Suitable for inside use only.	Quick.

All non-poisonous except lead pigments

Light reflectance approx 88% for all types.

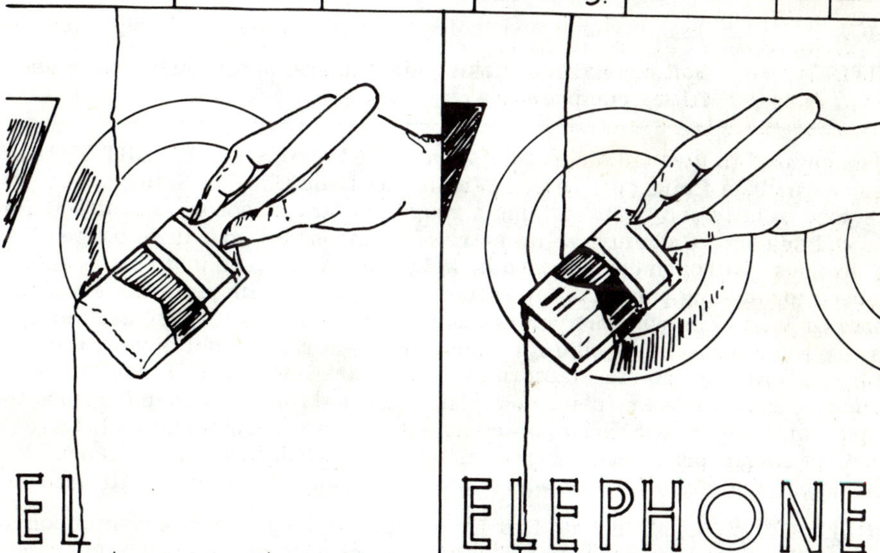

Opaque paint blots out the background.

Translucent paint fails to blot out the background

WHITE PIGMENTS • OPAQUE & TRANSLUCENT PAINTS.

Paints

EXTENDERS AND YELLOW AND RED PIGMENTS

All the foregoing white pigments are manufactured, but certain naturally occurring white minerals are used to cheapen the paint or give special properties, such as reinforcing the film or giving tooth or grip to it, and for making distempers and waterpaints. When used in oil paints these materials are known as extenders. The main ones are: BARYTES, which is very heavy and used to strengthen paint films and as a pigment for waterpaints; ASBESTINE, used as a suspending and flatting agent, being quite unaffected by moisture; CHINA CLAY, a fine soft and inert material used as a suspending agent, a base for organic dyes, a flatting agent in undercoatings and cheap oil paints, and making distempers; SILICA, a fine textured material used as a suspending agent and to give tooth or grip for flat wall paints, undercoats and white line road paints; WHITING, a soft crystalline material which dissolves in most acids and in rainwater. Being alkaline in reaction, it affects colours like Prussian Blue and Chrome yellow; it is mainly used for distempers, waterpaints, putty and filling compositions. Barium Carbonate, Strontium White and Gypsum are also used. A suspending agent thickens the oil without changing its colour and prevents heavy pigments settling on the bottom of the can.

MINERAL	CHARACTER AND USES
BARYTES	Heavy, strengthening paint films, waterpaints.
ASBESTINE	Light, unaffected by moisture, suspending and flatting agent.
CHINA CLAY	Soft and inert, suspending agent, base for dyes. Distempers.
SILICA	Fine texture, suspending agent gives tooth and grip, road paints.
WHITING	Soft crystalline, making distempers, water paints, putty and filling compositions.

Pigments used to give colour to paints are called 'stainers'. BRIGHT YELLOW pigments include Primrose, Lemon, Middle and Deep Chrome Yellows. All have good opacity and great staining strength but are poisonous and affected by lime and sulphur. Cadmium yellows are non-poisonous and a little deeper than the chromes. Although decomposed by acids, they are alkali and heat resistant and used for stoving paints. They should not be mixed with pigments containing copper or lead. Zinc chrome yellows have a greenish tinge. They are semi-transparent, soluble in all mineral acids, and used in anti-corrosive primings for light alloys and the manufacture of green paints. DULL YELLOW pigments include Ochres, which are permanent, fairly good stainers and semi-opaque in oil and water, and Raw Siennas, which have the same characteristics but are deeper in colour and more transparent. ORANGE PIGMENTS are Orange Chrome and Cadmium Orange, which are similar to the yellows in all but colour.

BRIGHT RED pigments include Red Lead, which is bright orange-red in colour, very similar to orange vermilion. Its value lies in its exceptional drying and protective qualities and is mainly used for primers. It is extremely heavy, poisonous, and darkened by sulphur. Vermilion is an orange-red, a very heavy pigment having fair body when ground in oil or water. It should not be mixed with lead or copper pigments. Nearest substitutes for vermilion are fast and signal reds and vermilionette. Scarlet Lake, Permanent Scarlet and Para Reds are all medium-toned reds. Scarlet Lake is somewhat transparent, but the other two are opaque. Crimson Lake and Permanent Crimson are bluish reds, the lake being the better colour, but transparent and not so permanent.

ORIGIN.	WHITE.	YELLOW.	ORANGE	RED.
LEAD.	white lead. basic sulphate of lead.	chrome yellow.	litharge orange. lead chrome.	red lead.
IRON.		yellow ochre raw sienna.	burnt sienna	venetian red tuscan red indian red.
ZINC.	zinc oxide zinc sulphide lithopone.	zinc chromate		
CHROMIUM.		lead & zinc chrome yellow.	orange chrome yellow.	
TITANIUM.	titanium di.oxide titanium white.			
COPPER.				
MERCURY.				vermilion.
COBALT.				
ANTIMONY.	antimony white.			
ALUMINIUM.	aluminium oxide china clay.			burnt island red.
BARIUM.	blanc fixe. barytes. lithopone.			
SULPHUR	basic sulphates of lead. barytes.	cadmium yellow.	cadmium orange	vermilion. cadmium red.
CALCIUM.	calcium carbonate calcium sulphate.			
STRONTIUM.	strontium white.			
CADMIUM.		cadmium yellow.	cadmium orange.	cadmium red.
SILICA.	asbestine silica. china clay.			
CARBON.				
COAL TAR. DERIVATIVES.		hansa yellow	orange II or orange lake.	fast red. para red. permanent red.

GROUPING OF PIGMENTS BY ORIGIN & COLOUR I.

Paints

BROWNISH-RED, BROWN, BLUE, GREEN AND BLACK PIGMENTS

BROWNISH-REDS include Venetian, Indian and Tuscan Reds, which are varieties of iron oxide. All are permanent in oil and water, have good opacity, can be mixed with any other pigments, and are non-poisonous. Burnt Sienna is similar but transparent. The main BROWN pigments are Raw and Burnt Umber, which are opaque, non-poisonous, permanent, suitable for oil and water media, and good stainers. The Raw Umbers have a greenish cast and the Burnt Umbers a reddish one. Vandyke Brown is very much like Burnt Umber but the organic matter it contains gives it a richer colour and renders it transparent. A poor drier when ground in oil, it is used for graining purposes ground in water and bound with glue size, vinegar or beer. BLUE pigments are as follows: Cobalt Blue, a light blue, somewhat expensive and only used where special colour effects are required. It is very permanent and works better in water than in oil. Ultramarine Blue is a bright blue, very suitable for distempers and waterpaints, as it is alkali-resisting. When used in oil it becomes transparent. Prussian Blue is a deep navy blue of exceptional strength, although somewhat transparent. It has good resistance to acids but the colour is completely destroyed by alkalis, and it is therefore not suitable for distempers. Monastral Blue is very similar to Prussian Blue but is not quite so deep in tone. It is absolutely permanent resisting the action of light, acids and alkalis, and can be used for all types of painting materials.

The principal GREEN pigments are lead chrome greens, zinc chrome greens and Chromium Oxide greens. Lead Chrome greens are mixtures of Prussian Blue and Lemon Chrome, which are sometimes called Brunswick Greens. All have good weathering properties, but the blue portions are affected by alkali and the yellow by acid. They have good opacity but are poisonous. The addition of black and orange chrome to these mixtures produces bronze and quaker greens. Zinc Chrome greens are mixtures of zinc chrome and Prussian Blue. They are non-poisonous but not so opaque as the lead chrome greens. Oxide of chromium green is a dull green, absolutely permanent and non-poisonous. Being unaffected by heat, acids and alkalis, it can be used for stoving paints, washable distempers and cement paints. Viridian has the same qualities as oxide of chromium green but is transparent and a pure blue-green in colour. The very poisonous Emerald Green has been replaced by Brilliant, Fast and Lime green. All BLACK pigments are light in weight and have good opacity and the majority of them are difficult to mix in oil and in water. Gas carbon blacks are intense in colour and have good staining power. Vegetable and lamp blacks are the high and low grades respectively of the same product and, although not so intense as the carbon blacks, they produce good bluish greys. Bone blacks are easier to mix and are preferable for mixing greys, as they are less liable to 'float' than the other types of black.

ORIGIN.	BROWN	GREEN.	BLUE & PURPLE	BLACK.
LEAD.				
IRON.	raw umber. burnt umber. vandyke brown.	terre verte.	prussian blue.	micaceous iron oxide.
ZINC.				
CHROMIUM.		viridian. oxide of chromium green.		
TITANIUM.				ilmenite black.
COPPER.		emerald green.	phthalocyanine blue. (monastral).	
MERCURY.				
COBALT.		cobalt green	cobalt blue cobalt violet.	
ANTIMONY.				
ALUMINIUM.				
BARIUM.				
SULPHUR.			ultramarine blue. (complex composition)	
CALCIUM.				
STRONTIUM.				
CADMIUM.				
SILICA.		terre verte.		
CARBON.				lamp black carbon black. graphite. vegetable black
COAL TAR. DERIVATIVES.		pigment green. brilliant green.	monastral blue.	

GROUPING OF PIGMENTS BY ORIGIN & COLOUR II.

Paints

DRYING AND TREATED OILS

The most important part of any paint is the binder or medium, because, if it fails to dry or harden properly, the paint rubs off easily or gets washed away. The materials used consist of drying oils both refined and cooked; treated semi-drying oils; and resins both natural and manufactured. These materials when spread thinly and evenly over a surface change from a fluid to a hard film which forms the actual protective coating.

Linseed oil is still the most important oil. The refined oil has a pale yellow colour and a mild and pleasant odour. It forms a glossy film in 3-4 days. Alkali refined oil is used for varnish making and acid refined oil for paints. Blown linseed oil, in which air is blown through heated oil, dries in 2-4 hours and is used for flat wall finishes. It varies in colour from golden yellow to red. Boiled linseed oil, in which drying agents are added during heating, are mainly of two types; pale, which dries in 24-48 hours, and dark, which dries in 16-30 hours. Boiled linseed oils vary from pale to dark brown in colour and produce films which are more glossy than those produced with refined oil. Stand linseed oil is a heat treated oil without any driers added; it dries in 8-12 hours and forms films which are paler, thicker and more durable than the other types. It is used for making enamels, gloss paints and printer's inks.

Perilla oil is similar in most respects to a mixture of refined linseed oil and tung oil. It dries in 2-2½ days. Tung or China wood oil has better water and alkali resisting properties than linseed oil. Refined tung oil dries in 2 days but the heat-treated varieties in 4-8 hours. It is deeper in colour than linseed oil but on storage the colour bleaches and it becomes wax-like. It has a musty smell. The refined oil dries without gloss and wrinkles badly, but the prepared varieties produce excellent media for paints, varnishes and cement and plaster primers. Oiticica oil is very similar to tung oil and dries in 1½-2 days; it is best when used with synthetic resins.

Of the treated semidrying oils, castor oil, mineral oil and rosin oil are the most useful. Treated castor oil produces a material which is similar to a mixture of linseed and tung oil and dries in 5 hours. It is used for varnishes, stoving paints and with synthetic resins. Mineral oil dries in about 2 days; being perfectly alkali resisting, it is chiefly used for cement and plaster primers. Rosin oil is used in the manufacture of emulsions and waterproofers. Tobacco seed oil is largely gaining in favour as a partial substitute for linseed oil, owing to its cheapness and non-yellowing properties, although it takes longer to dry. Other oils used in paint manufacture, either mixed with drying oils or alone are fish oil, which is used for stove enamels and floor paints, safflower oil as a substitute for linseed oil, soya bean and tall oil.

PAINT MEDIA MUST WITHSTAND FLUCTUATIONS OF TEMPERATURE, ACIDS, STEAM AND ALKALIS.

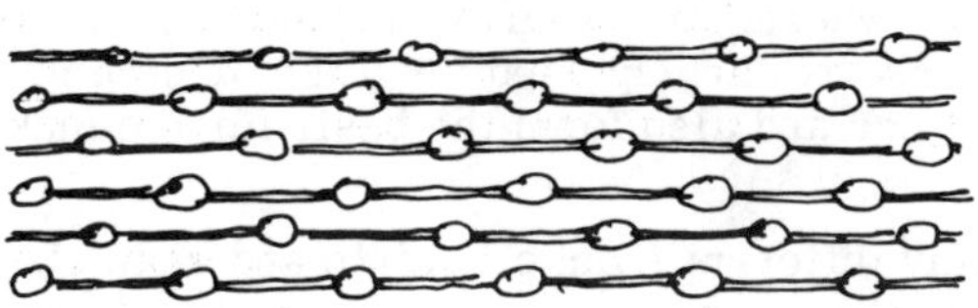

The bonds of the oil form a film and cement the pigments on the surface.

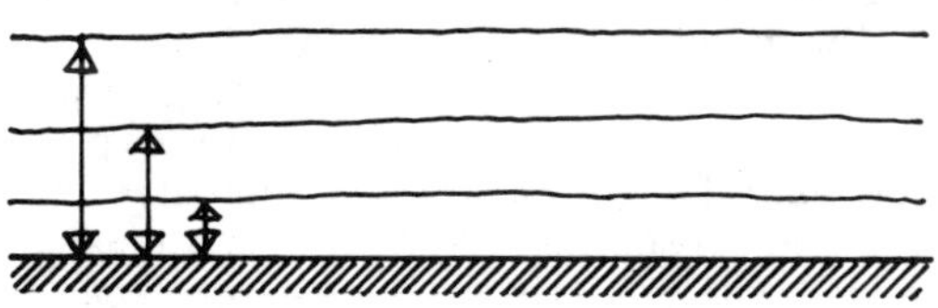

It is best to apply thin coats so that the media hardens right through.

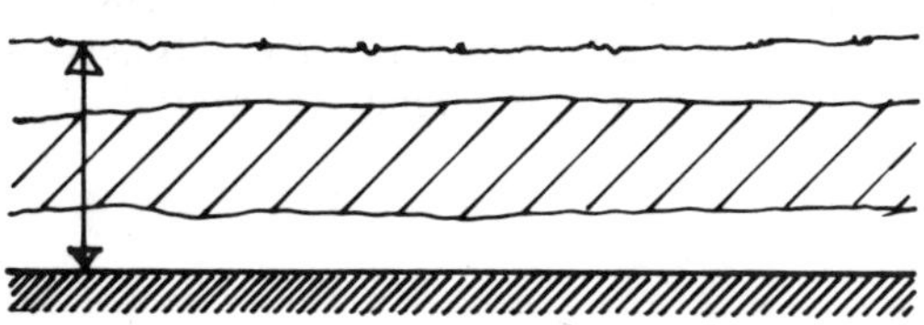

If the coat of paint is too thick, the top skins over and cuts off the air from the centre, and it remains soft and wrinkles.

WHAT PAINTS ARE EXPECTED TO DO.

Paints

NATURAL AND SYNTHETIC RESINS

Resins are used to reinforce and strengthen the oil films and give additional gloss. These oil and resin mixtures are referred to as short, medium and long oil paints depending on the proportion of oil they contain. They are also used for making varnishes; and the varnish or medium obtained depends upon the proportions of oil and resin. The larger the quantity of oil, the more elastic the material will be. Natural resins are either exudations from living trees, which are called copals, or fossil deposits of forests. Copals are traded as varieties from their country of origin: e.g. Zaïre, Sierra Leone and Tanzania, and are hard and brittle resins. Other natural resins are dammar, used for oil and spirit varnishes and cellulose ester products, kauri for oil varnishes, mastic for high class paper varnishes, and rosin. Rosin is the residue left from the extraction of turpentine; it is brittle, highly acid and needs neutralizing with lime or zinc oxide to make it miscible with oil. Treated with glycerin, it produces ester gums which are plastic, resistant to weather, free from 'feeding' (thickening) with basic pigments, and have good alkali resistance. Ester gums are used extensively with tung oil and cellulose. Natural resins used for spirit varnishes are shellac, manila and dammar; the latter which dissolves readily in white spirit, is used for paper varnish.

Manufactured or synthetic resins include phenolic, alkyd, maleic, urea, epoxy, acrylic, melamine, vinyl and polystyrene. Alkyds are the most widely used, different varieties being employed for the majority of painting materials, depending on the amount of oil they contain. Long oil types with 60-70 per cent of oil are suitable for brushing qualities of all types; medium oil types containing 50-60 per cent oil for brushing, spraying qualities and air drying industrial finishes; and short oil with only 42-50 per cent oil for spraying and dipping materials. Epoxy resins are used where hard-wearing and anti-corrosive properties are essential. Vinyl resins are largely used for wash or etch primers, for both two pack and self etch types and also form the basis for the making of plastic emulsions.

CELLULOSE PAINTS are entirely different from oil paints and enamels as they do not undergo any chemical change during drying. There are many formulae for making lacquers and enamels but all are based upon 'nitrocellulose' dissolved in ether-alcohol. Owing to their rapid drying, the only means of covering large areas is by spraying. In painting and decorating, their chief use is for coating interior plaster surfaces. All materials of this type are highly inflammable. BITUMINOUS PAINTS are made from asphalt, bitumen or pitch, which are soluble in naphtha and turpentine. They are either used alone to form Brunswick blacks or blended with rosin to produce black japan and stove enamels. Most bituminous paints are sensitive to light and changes of temperature. Neutralized tar paints offer excellent protection against chemical action and the combined action of wind and salt water, and are extensively used for marine work. Most tar products have a tendency to 'bleed'.

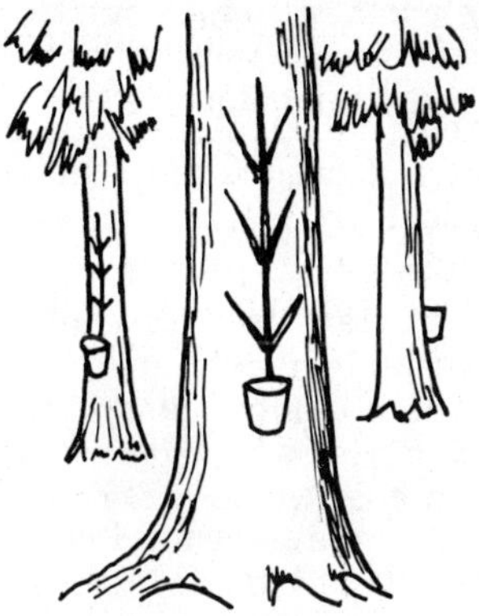

Natural resins are either exudations from living trees (Copals) or are fossils dug from the earth.

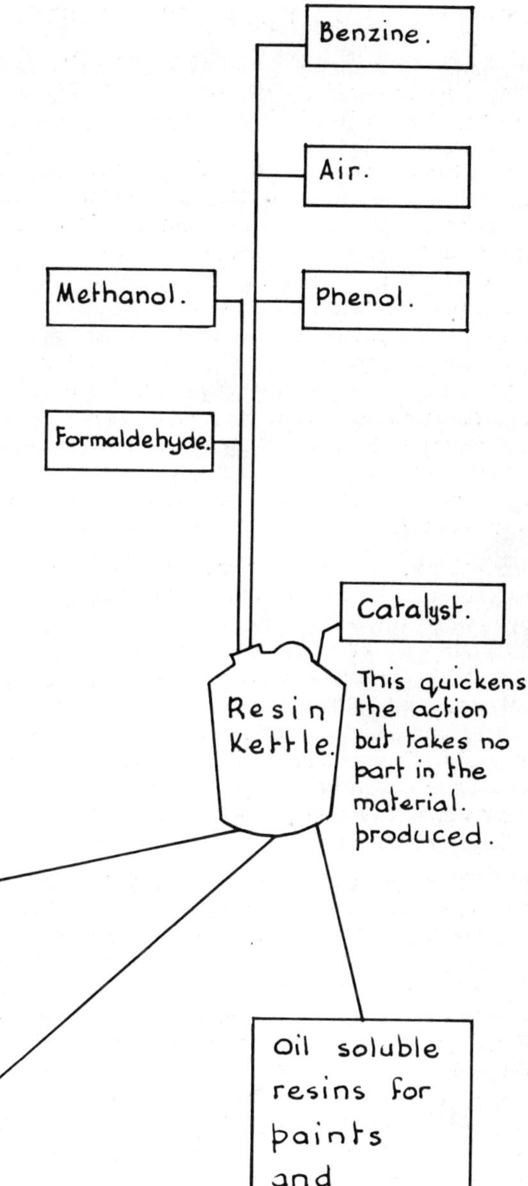

Synthetic resins are manufactured.

FLOW SHEET SHOWING HOW PHENOLIC RESINS ARE PRODUCED.

SOURCES OF NATURAL AND SYNTHETIC RESINS.

DRIERS AND THINNERS

Most ready mixed paints are quick drying to enable another coat of paint to be applied with the minimum of delay and ensure that the finishing coat does not get soiled while it dries. As painting has to be done in all weathers and conditions, some of which are anti-drying, like frost, cold, damp and fog, it is necessary to add materials which will help and ensure the drying of the paint. Most driers consist of salts of lead, manganese and cobalt mixed together in certain proportions according to the type of paint. The lead salts are litharge, red lead, acetate and borate of lead, all of which are good driers but darkened by sulphur. Manganese compounds are more powerful than lead but not so regular in action; they tend to darken pale colours giving them a pinkish cast. Cobalt compounds are the most powerful and only small quantities are required, but paints containing them are liable to dry too slowly in cold weather and too quickly in hot. The latest type of drier is naphthenates, which contain certain proportions of lead, manganese and cobalt, and are very reliable and stable. Materials used by the painter on the job for speeding up the drying of paints include liquid driers, terebine, japan gold size and mixing varnish. Paste driers can also be obtained but their use is limited. Driers must always be added in very small quantities (about one teaspoonful to a pint), otherwise the drying of the paint may be retarded and eventually cause it to crack.

If a material is too stiff and cannot be spread smoothly over a surface, it will not be satisfactory and must be thinned, i.e. have its consistency reduced. This is all-important as paint which is too thick will not dry throughout its thickness and may eventually cause the paint film to wrinkle, rivel or crack. Paints which are hard to apply and cannot be 'evened' out satisfactorily lead to overloading of the surface, coarse brushwork and the use of more paint than is necessary to obtain a satisfactory finish. A thinner should be colourless, not affect the material to which it is added and evaporate without leaving any residue in the film. Some paints require special thinners and this is generally indicated on the container; but most oil paints, undercoatings and gloss paints can be thinned with white spirit. Only the best quality of white spirit should be used, as the poorer qualities tend to leave a greasy residue or 'tail' in the paint film, which escapes at a later date and causes loss of gloss and, in severe cases, pitting of the paint film. Oil varnishes, including japan gold size are best thinned with turpentine, as there is less likelihood of the resins being thrown out of solution and curdling; but the best qualities of white spirit are also useful. For knotting and spirit varnishes, methylated spirit should be used. Bituminous products require solvent naphtha, tar spirit or turpentine, and cellulose needs amyl or butyl acetate, and nitro-paraffins. All these materials slightly soften the underneath coat so that the material being applied grips the surface and forms a single uniform coating. Washable distempers and waterpaints can be thinned with petrifying liquid or water, according to the surface being treated. Emulsion paints, wool grease emulsions, cement paints and soft distempers need only water. Thinning enables a thinner film to be applied and in some measure assists drying.

Paint that is too thick.
will soon peel.

PAINTS NEEDING DRIERS.	PAINTS NOT NEEDING DRIERS.
gloss paints.	red and white lead primings.
oily undercoats	red lead and red oxide paints
glazes.	spirit paints.
oil paint finishes.	cellulose paints.
oil graining colour.	
zinc and titanium based paints.	

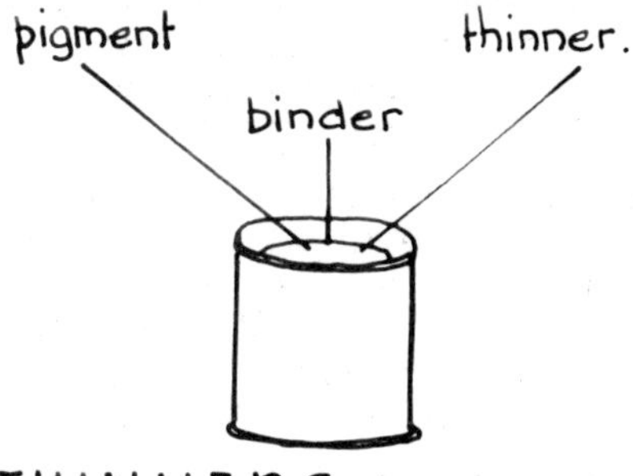

DRIERS AND THINNERS.

Specialized Paints

PRIMERS AND INDUSTRIAL FINISHES FOR METALWORK FUNGICIDAL PAINTS, INSECTICIDAL PAINTS, WOOD PRESERVATIVE SOLUTIONS

Primers and industrial finishes for metalwork, which include vehicle chassis paints, machinery finishes, enamels for toys and general maintenance paints are of three main types—styrenated alkyds, toluenated alkyds and chlorinated rubber. (1) Styrenated alkyds which have a medium of oil modified alkyd resin combined with styrene (a benzene derivative) set so quickly that brush application is impossible and the only satisfactory methods are dipping and spraying. They are less likely to wrinkle than other types of finish, and have good resistance to water, chemicals, humidity and bloom. If the percentage of styrene is high they are less sensitive to white spirit, and stronger solvents such as xylenes and toluene are required. (2) Toluenated alkyds, resin media in which toluene is used instead of styrene, can be thinned with white spirit and produce excellent primers with red lead and zinc chromate as the pigment. They can be brushed or sprayed as required. (3) Chlorinated rubber based paints which have excellent resistance to mild chemicals, moisture and mineral oils but do not stand up to severe contamination of vegetable and animal fats.

The latest development in paints for metals are acrylic resin based primers which can be water thinned. In the car industry this cuts down the fire risk and cost, but satisfactory techniques have still to be developed when damaged surfaces have to be stripped and touched up.

Fungicidal paints are necessary in premises used for the storage or processing of food and beverages, such as bakeries, breweries, dairies and banana stores, to combat the growth of mould or fungi.

Before any paint is applied it is essential that all surfaces should be thoroughly cleaned and all existing mould growths killed. Fungicidal washes such as Shirlan N.A. or Santobrite are suitable and two applications should be given. The paints used should dry hard and with a full gloss to prevent lodgement of spores; and paints containing chlorine derivates such as chlor-phenol, zinc oxychloride or chlorinated rubber are strongly recommended. Where there is no risk of food contamination paints containing mercuric chloride should be used as they are more effective.

Where dry rot has to be eradicated it is essential to remove and burn immediately all affected materials to a distance of 1·25 m beyond the affected area. Brickwork, stonework and ironwork should first be treated with the flame from a blow lamp or torch, and all surfaces given two applications of sodium fluoride or magnesium silico-fluoride (approximately 40-50 g dissolved in 1 litre of water). As an added precaution all new timber should be treated with the fungicide before being fixed.

Insecticidal paints are recommended for kitchens, canteens, hospitals, warehouses, ships and all premises where insect infestation can be harmful to health and hygiene. They are similar in appearance to clear varnishes and ordinary gloss paints, and the preparation of the surface to receive them and the method of application are identical. Washing the painted surface generally re-activates its insecticidal properties.

Wood preservative solutions are generally based on pentachlorophenol for protection against fungicidal attack, and polychloronaphthalene for the eradication and prevention of attack by wood boring insects. Timbers treated with these solutions can be subsequently varnished, painted or polished.

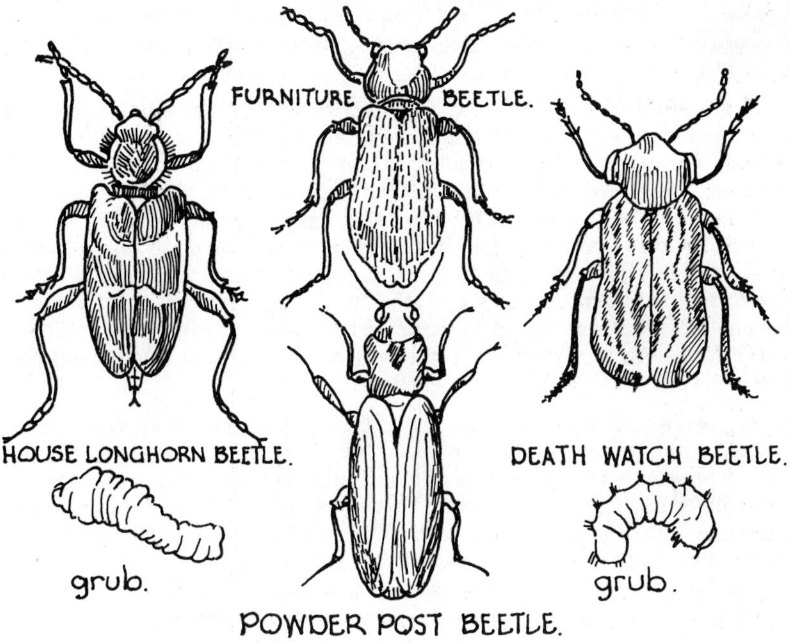

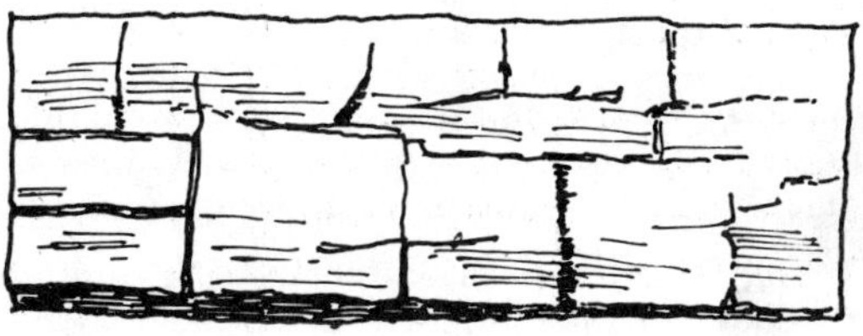

TYPICAL CUBOIDAL EFFECT OF DRY ROT UPON TIMBER.

DRY ROT - A RIPE SPOROPHORE.

WOOD BORING INSECTS & FUNGAL ROT.

Specialized Paints

EPOXY RESIN BASED PAINTS

These will generally withstand conditions which are too severe for conventional gloss paints. The media for all such paints are modifications of imported resin. There are three main types:- epoxy esters, epoxy-phenolic, and epoxy-amino paints.

Esters are made by reacting a fatty acid with an alcohol e.g. linseed oil and glycerine, and with epoxy esters the resin as an alcohol is reacted with the fatty acids of linseed or de-hydrated castor oil to produce media similar to alkyd resins. The percentages of oil and resin used in their manufacture govern their use and behaviour. Those with large percentages of oil (long oil esters) are used for priming and decorative paints on wood, metal, stone, brick and plaster. They give good resistance to dilute acids, and alkalis, detergent solutions, chemical fumes and excellent corrosion resistance under severe conditions. Application can be by brush or spray and white spirit used for thinning them. Although touch dry in 4-6 hours, maximum chemical resistance does not develop until 48 hours after drying.

'Cold cure' esters are of the 'two pack' type, supplied in two containers—a paint base in one and a catalyst in the other. The latter is a substance which hastens a chemical or physical reaction (such as the curing or drying of a paint) but does not itself enter into the reaction. The two materials are mixed together immediately before use, and once mixed, can only be used during the next 24 hours. Application is by brush or spray and special thinners are required. Although touch dry in two hours, recoating should not be attempted before 24 hours. Maximum adhesion, flexibility and resistance develop seven days after drying at room temperature but 'curing' is retarded by cold conditions. These materials give good resistance to acids, alkalis, solvents, oils, detergents and hydraulic fluids.

Short oil esters combined with urea and melamine-formaldehyde resins are used for stoving finishes for cars, refrigerators and other machinery. The stoving times are usually 15-30 minutes at approximately 120°C.

Epoxy-phenolic systems give the greatest chemical resistance of any known surface coating but generally the films are brittle and poor in colour. The usual application is by spraying or dipping and afterwards stoving at 200°C. for 20-30 minutes. A catalyst such as phosphoric acid is generally added to hasten the hardening.

They are mostly used for the interior coating of pipes for the transporting of chemicals.

Epoxy-amino systems are combinations of epoxide resins with either urea or melamine-formaldehyde resin or a mixture of both. They can be stoved at much lower temperatures than the epoxy-phenolics and there is very little discolouration of white and pastel shades. They have extremely good flexibility and adhesion but not the same chemical and physical resistance as the epoxy-phenolic paints. Owing to the high cost of epoxy resins it is not unusual to find in epoxy-urea systems part of the epoxy resin replaced by an alkyd resin.

'TWO PACK' MATERIALS, MULTI-COLOURED SPECKLED FINISHES, FIRE RETARDING MATERIALS

With two pack materials the base paint will dry in the same way as an ordinary paint will, so if the addition of the contents of the second can is overlooked, no great harm will be done. The contents of the second can constitute a hardening agent which will give to the paint an added quality of extra chemical resistance. This extra hardening is known as 'curing'. For this reason only enough paint base and hardener should be added together that can be used during the one working period. Hardeners are usually based on sulphuric, hydrochloric or phosphoric acid.

Wood finishes and neoprene based rubber paints are also 'two pack' materials. With wood finishes the ester is blended with urea resin and cured at room temperature by the addition of an acid catalyst such as hydrochloric acid. They dry rapidly and the surface can be polished and buffed within 3-4 hours. Rubber paints based on neoprene are highly flexible but black in colour. Under normal temperatures they have fair adhesion to most surfaces but on metals and where high temperatures are met with, bonding primers must be used. These materials are termed 'high build' as they have a high solids content, so that a film 0·2 mm thick can be applied with each brush application. They have good resistance to abrasion and very good chemical resistance against acids, effluent and sea water.

Making a speckled surface by spraying flecks of different colours on a painted ground has always been popular, but whereas formerly each colour had to be applied separately, materials recently introduced enable the ground and spots of three or four different colours to be sprayed in one operation. Based on two media that do not mix (cellulose and water) they dry in about two hours, and are hard dry in four to five hours. They can only be applied by spray with a low pressure gravity feed gun. If faulting occurs, any touching up can be done without showing, and the finished surface is tough and hard wearing. It is claimed that these materials do not produce static electricity and consequently dirt will not adhere so readily as it does to ordinary paint finishes.

Fire retarding materials are often used in building. For treating timber prior to construction, pressure impregnation or hot and cold dipping with a mixture of ammonia and boric acid is effective; but the hygroscopic (moisture attracting) properties of some of these materials may warp the timber or promote corrosion of any metal in contact with timber so treated.

Flame restricting materials applied to existing buildings are of three types: (1) Those which swell and froth (tumefy) under the action of heat and form a hard insulating layer on the surface. Sodium silicate, and also combinations of urea-formaldehyde resins and ammonium phosphate behave in this way. (2) Those which give off a fire retarding gas (chlorine) and consist of antimony oxide mixed with chlorinated rubber or chlorinated paraffin. (3) Thick coatings of materials which tend to insulate the surface from the heat such as asbestos, borax and kaolin mixed with a suitable binder, since these materials have low thermal conductivity (i.e. do not allow heat to pass through them quickly); or calcium sulphate anhydrite plasters which until all the water of crystallization is driven off will not allow the temperature of the surface to rise to the point of ignition.

Specialized Paints

TREATING DAMP SURFACES, THIXOTROPIC PAINTS, NON-YELLOWING PAINTS AND STOVING ENAMELS

When paint is applied to a damp surface, there is always a risk of it peeling or blistering, and it is often extremely difficult to assess the amount of time necessary for a surface to become completely dry or how soon it is safe to apply paint.

For testing the quantity of moisture in a surface two small instruments are available: (1) A hair hygrometer, mounted in a case, which can be fastened on the surface with putty or plasticine and left for an hour or so. The hair, fixed to a pointer, expands or contracts with the degree of humidity and so registers the moisture content. (2) A battery operated instrument with two leads to hard metal points which when pressed in the surface immediately registers the moisture content on a graduated scale.

Special paints, termed hydro-treated, are manufactured to reduce the risk of lack of adhesion due to moisture. The active agent is amine stearate solution which is incorporated in the paint so that when applied by brush, the paint strikes deeply into the pores of the surface in spite of the moisture.

If the viscosity (degree of thickness) of a paint is reduced or it becomes thinner when stirred or brushed but thickens again when the movements cease it is said to be 'thixotropic'. Earlier types were inclined to have poor flow and inferior drying but the latest have 'built in thixotropy' in the alkyd resins used in their manufacture which makes them easy to apply, enables fuller coats to be applied without runs or sagging, and prevents the pigments settling in the container. These materials have little or no smell and dry in two or three hours. They are ideal where flat or eggshell finishes are required and for silk screen printing.

Painted surfaces are discoloured by the action of bright light, diffused light, darkness and heat. Bright light turns light coloured paint yellowish, because ultraviolet radiation attacks the oil of the medium and causes coloured compounds to be formed. This discolouration is, to some extent, lightened by being bleached by other ultra violet rays. Diffused light or darkness act in the same way, but because of the lack of ultra violet rays, the yellowing becomes permanent.

Heat also causes yellowish coloured compounds to be formed which cannot be bleached or destroyed by ultraviolet radiation. To overcome this, paint made with a medium of soya bean or tobacco seed oil is used. For non-yellowing stoving enamels, either urea or melamine resins are combined with alkyd resins; the proportion is generally $1\frac{1}{2}$ times the amount of alkyd to urea, or 4 times the amount of alkyd to melamine with coconut or castor oil. These materials have good gloss, flow and adhesion.

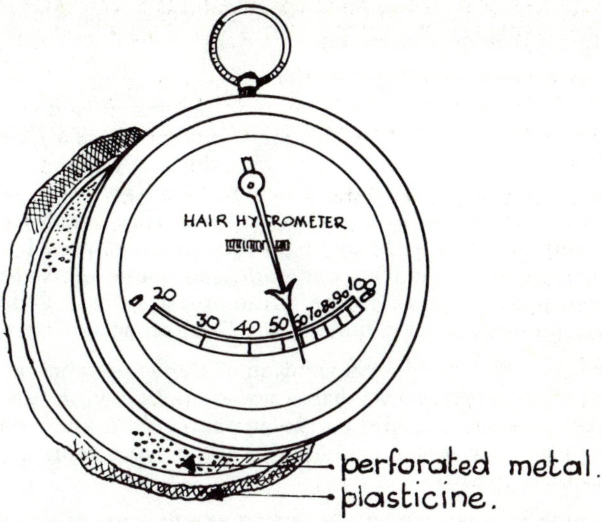

HAIR HYGROMETER FIXED ON WALL TO RECORD MOISTURE.

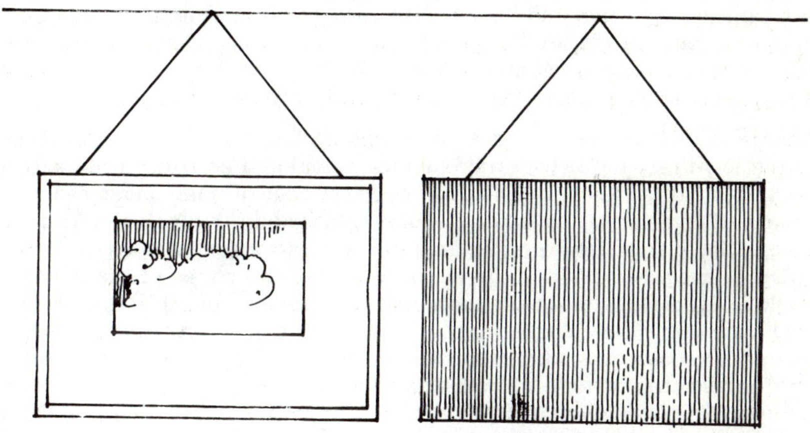

DISCOLOURATION LEFT BY PICTURE. ACTINIC RAYS WILL BLEACH OUT MARK TO A CERTAIN EXTENT EXCEPT IN DARK SITUATIONS.

SPECIALIZED PAINTS · HYDRO TREATED & NON-YELLOWING.

The Drying and Adhesion of Paints

The reactions that take place when painting materials change from fluids to dry films vary according to the type of material; and, while some reactions are simple, others are very complex.

Primers, undercoatings and finishes containing a large proportion of oil dry by absorbing oxygen from the air after the thinners have dispersed by evaporating, and this process is termed oxidation. Gloss paints and finishes containing less oil but a good deal of resin dry less by oxidation than by polymerization, a process in which the molecules of resin and oil group together to form polymers, which in turn cluster together until the film becomes solid. Thus, in each case the medium in the dry film differs from when it is in a fluid state, but polishes including knotting and cellulose materials which undergo no chemical change dry by the evaporation of the spirit solvent, and can easily be removed from the surface by rubbing with a cloth soaked in solvent.

Soft distempers dry by the evaporation of the water content and can easily be washed off with water, but washable waterpaints dry in two stages; first, the particles of pigment are held on the surface by the glue content, and then later the coatings become water resistant by the oil and resin oxidizing and polymerizing, as with oil paints.

Factors which prevent oil paints drying are grease, cleaning materials left on the surface, draughts, gas fumes, stagnant air and condensation. Lack of ventilation delays evaporation and prevents oxidation; condensation delays both evaporation and oxidation; and draughts increase evaporation but prevent oxidation. After painting windows and doors, it is better to open them as wide as circumstances permit rather than leave just a small gap through which the wind can race. But coatings which dry but do not adhere to the surface are useless and upon adhesion depends the success of all paint systems.

Adhesion is the attraction of two different surfaces in contact resulting in their sticking together, and before any surface is painted consideration should be given to its nature and condition, as the first coat of paint must react favourably and adhere firmly to form a sound foundation for all the paint afterwards applied. Materials like wood and plaster are porous to a varying degree and require primers which stop the suction and fill the pores of the surface with a sticky material to which the next coat will adhere and this is called mechanical adhesion. New plaster and cement surfaces contain substances which under damp conditions can affect the oil in the paint, and the primer selected should in addition be able to withstand such chemical reaction. With non-porous surfaces like tiles, glass and certain metals, reliance must be placed on the adhesion between the surface, the priming and undercoat and this is specific adhesion. Surfaces like iron and steel which rust quickly need primers that form a barrier to fluids and, if possible, prevent corrosion by stifling the corrosive substances, and, in some cases, withstand heat.

Surface. Coating.

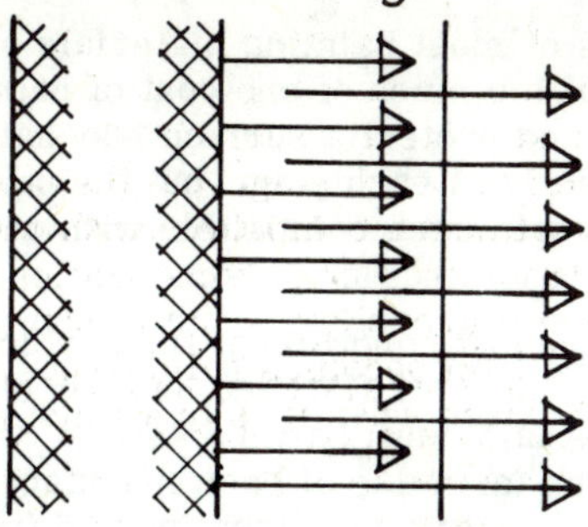

POLISHES, KNOTTING
CELLULOSE MATERIALS.

Surface. Film.

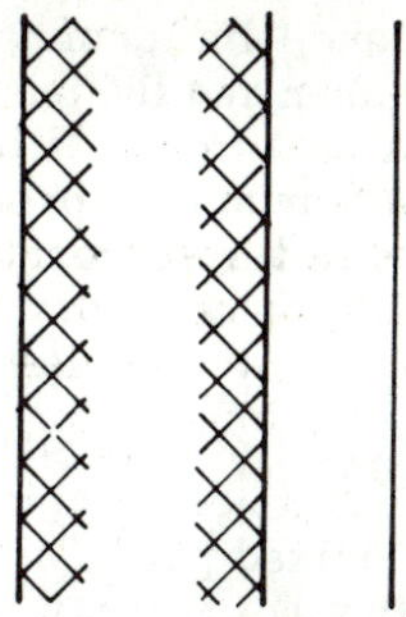

The solvents evaporate and leave a solid film on the surface.

OIL GLOSS PAINTS.

Surface.

① The thinners evaporate. (Evaporation.).

Surface.

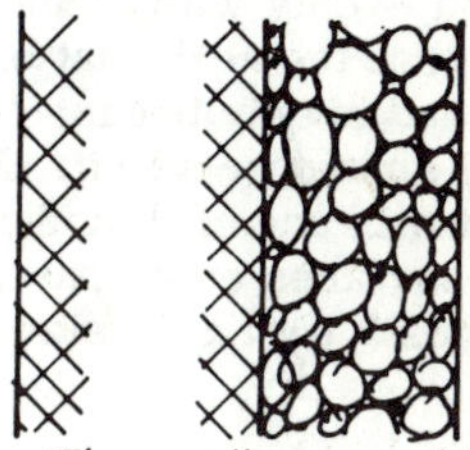

② The oils combine with oxygen. (Oxidation.)

③ The oxyns group together as the paint hardens. to form polymers and this is termed polymerization.

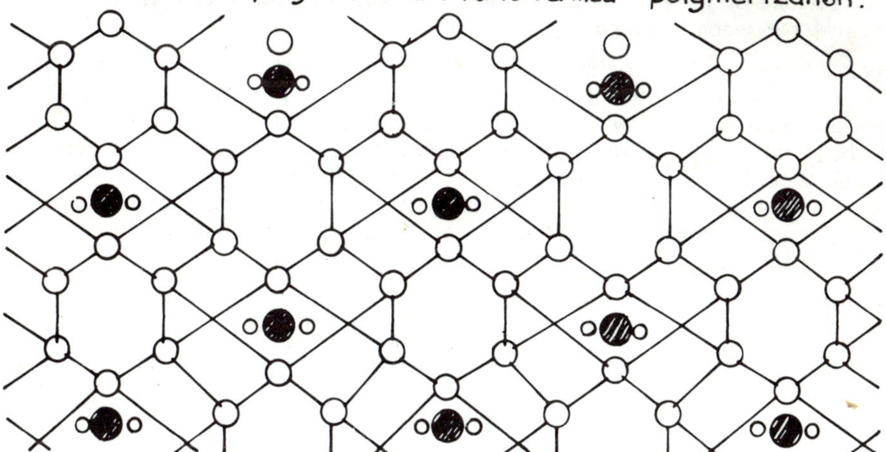

DRYING OF PAINTS.

Paint Systems

THE IMPORTANCE OF PRIMING COATS

The good drying and film forming properties of most painting materials often cause both craftsmen and the public to expect too much of one coat of paint, and even a very thick coat cannot fill, obliterate and protect a surface adequately. Not only must the surface be prepared correctly at each stage, but the paint system built up with a sufficient number of coats each compatible with one another.

Paint systems vary according to the substance and condition of the surface, type of material and finish required (whether gloss or flat) and number of coats specified, but general principles can be laid down based on sound craft practice. Ready-mixed paints are made to the formulae of each manufacturer and it is better (but not essential) to use the primers, undercoating and finishes of one particular manufacturer throughout a job rather than change to another during it. Methods of building up a paint system vary; some craftsmen prefer to build up the paint film with each succeeding coat of paint getting more elastic; others prefer alternate flat and oily coats.

The priming coat should contain sufficient oil, gold size or varnish to enable it to be applied evenly with brush, spray or roller, and to seal the surface sufficiently to prevent the next coat of paint 'sinking'. Ready-mixed primers which cannot be spread evenly because they are too thick should be thinned with a mixture of oil and white spirit. On very porous surfaces the paint should be thinned with a mixture of 2 or 3 parts linseed oil and 1 part white spirit; for semi-porous surfaces equal parts of linseed oil and white spirit; and for non-porous surfaces 1 part gold size or varnish and 2 parts white spirit or turpentine.

The colour of the priming coat greatly influences the colour effect of the finish obtained. For woodwork which is to be finished white or a very light colour, a white lead or very pale pink lead primer would be better than a deep pink or red lead one. Ironwork which is to be finished in a light colour would be better primed with aluminium or a buff coloured calcium plumbate primer than a darker red lead or grey primer; but the latter would be more effective for brown, grey or green finishes.

Between the drying of the priming coat and the application of the second coat, all necessary rubbing down, stopping and filling should be done. If when the primer has been rubbed down it still has a fair degree of sheen, the next coat should be a flat one; but if the rubbing produces a flat or dull surface a little oil can be added to the next coat. Surfaces which are extremely porous, such as very soft building boards, will need to be treated by coating with glue size solution, proprietary wall-board sealer, or a mixture of oil and white spirit before any paint is applied.

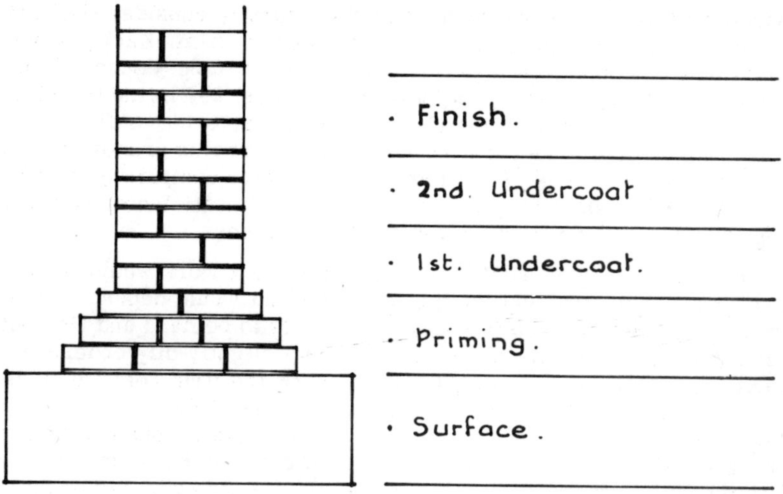

Priming coats are the foundation for all paint systems.

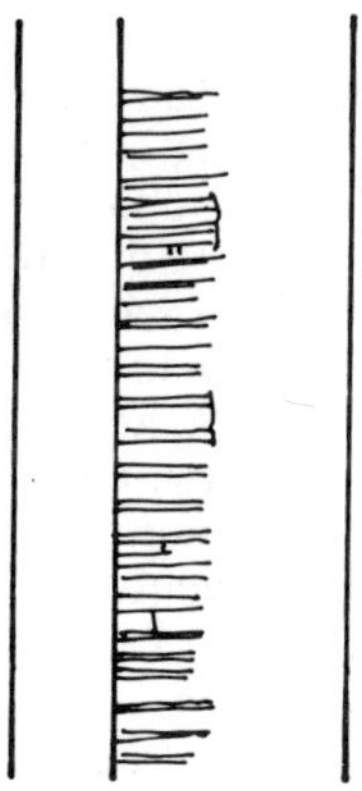

On porous surfaces some dovetailing of the priming paint and the surface takes place.

On non-porous surfaces the priming paint must rely only on adhesion.

THE IMPORTANCE OF PRIMING COATS.

Paint Systems

THREE AND FOUR COAT; USING DIFFERENT MATERIALS

To understand how paint systems work, it is necessary to consider the various types of paint that can be used to build up a satisfactory film. Each layer or coat of paint must be compatible with the underneath surface (especially if it is a priming coat) and the coat above. Undercoatings must key to the priming and dry hard enough to form a foundation for another undercoating or finishing coat to be applied. Mixtures of undercoating and finish in varying proportions produce materials which make excellent intermediate coats compatible with both undercoats and finish; and, although most materials mix satisfactorily, there are a few exceptions which are usually noted on the container.

With interior painting, a three coat paint system for new work would be priming coat, undercoat and finishing; and the type of undercoat would depend on the finish required. For a gloss finish the undercoat needs to be hard and flat, but for a flat finish the best results are obtained with a slightly oily undercoat. For a three coat system on previously painted work, the first coat should be flat or sharp, so that it keys to the paint already on the surface; the second coat undercoating or a mixture of equal parts of first coat and finish; and the last coat the gloss finish. For a flat finish, the second coat can be mixed with 3 parts flat finish and 1 part linseed oil. This would ensure evenness of the colour finish and prevent three coats of paint of a similar nature, i.e. of small oil content, being applied.

Four coat systems, gloss finish on new work including the priming, can be built up in two ways; for hard gloss paints two undercoats and one coat of finish are usual, but with synthetic gloss finishes only one undercoat and two coats of finish are advised. An alternative method would be to mix equal parts of undercoat and finish for the coat before the finish in each case.

Where systems are built up with different materials, such as a bitumen solution over a red lead primer, it is essential that the priming coat is thoroughly dry and hard before the next coat is applied. Oily coats can be applied over flat coats quite safely, but very quick-drying coats applied over oily ones cause the paint film to crack. Oily coats applied over oily coats are never satisfactory. For a grained finish on new work, the usual practice is to prime, ground in for graining, grain and afterwards varnish. Where the graining is in water-colour, gloss varnish will be needed, but if in oil the varnish can be flat. For cheap work the graining is usually left unprotected. Satisfactory paint films can only be built up with coats of paint that are applied evenly. Uneven application causes stresses to be set up which tend to break down in the valleys of the brush marks.

Sealers should not be confused with primers. Their chief function is to correct or overcome excessive suction of the surface and enable the primer to be applied easily and economically. Quite different types of sealers are used to prevent coloured materials bleeding through freshly applied paint. These can be spirit varnish or knotting, aluminium wood primer or even black paint.

1 Systems of building up paint film with alternate fat and lean coats.

2. Each coat becoming progressively more elastic.

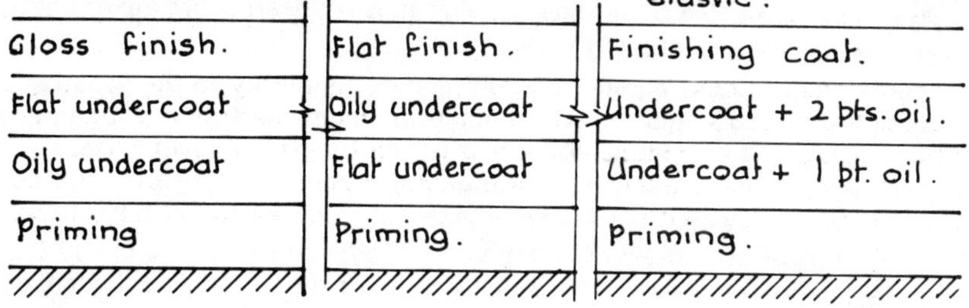

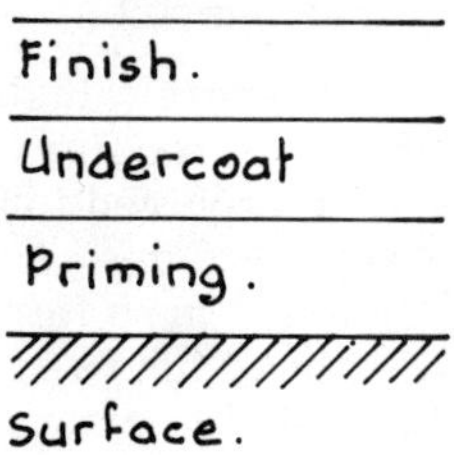

3 coat paint system.

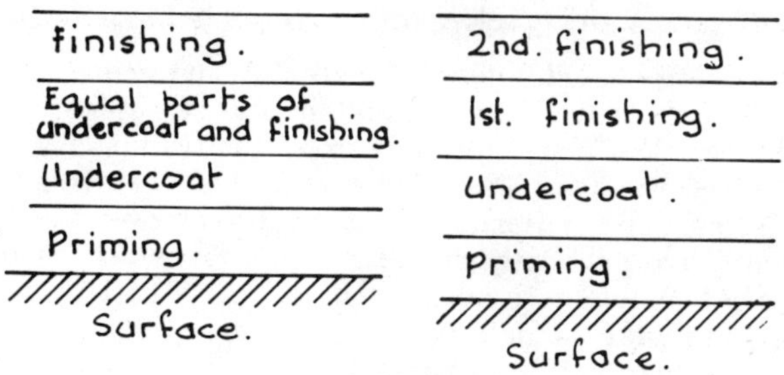

Two methods of building up a 4 coat paint system.

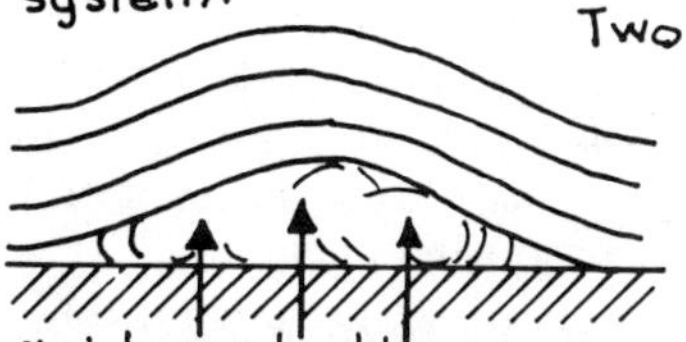

Moisture trapped under the paint film may expand and cause blistering.

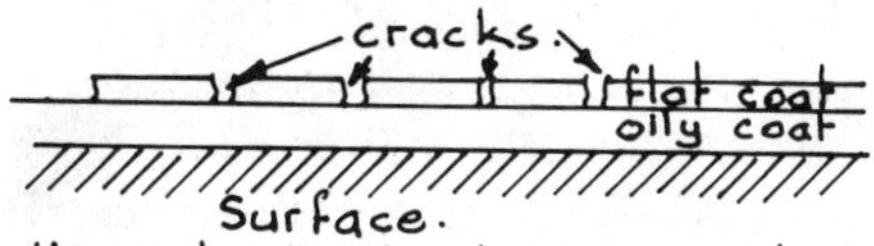

Unequal expansion between coats, such as a sharp flat paint over an oily undercoat may cause cracking of the paint film.

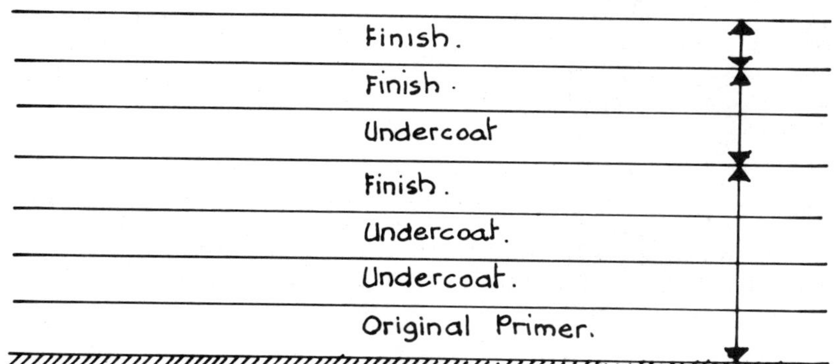

Overloading of the surface may cause peeling or flaking.

PAINT SYSTEMS.

GLOSS PAINTS AND THEIR APPLICATION

Although there are many types of paint marketed, they either dry with a full gloss, or are flat with little or no sheen. The main types of full gloss paints are oil gloss, enamels and synthetic paints.

Oil gloss paints are known variously as high gloss, hard gloss, enamel paints or gloss paints, and their chief characteristic is that they flow out to a smooth film without brushmarks. The best qualities dry hard, have good adhesion, brilliance, retain their gloss and have reasonable opacity; and these qualities combine to make them wear and weather resistant. Gloss paints should be applied over a hard, flat and solid undercoat. They are usually brushed on, but can also be sprayed. Brushing should be firm, with the bristles going through the paint, and it should be laid off in one direction. Brushing should continue until there is a slight drag on the brush. Drying times vary, being generally from 6-20 hours, the quicker drying types being generally less durable.

Enamels are a superior type of material, and are chiefly used where a high gloss and durable first class finish or where periodic cleaning is required. Two qualities are made and it is important to specify whether required for inside or outside use. They flow better than the cheaper types of gloss paints and consequently need brushing out to a very even film. Enamels have fair opacity and are generally dust-free in 8 hours and hard-dry in 24 hours. If gloss paints or enamels need thinning, only a little turpentine or the manufacturer's thinners should be used. Under normal conditions the spreading capacity of gloss paints is about 13-15 m^2/litre. Enamels will cover a few square metres more.

The undercoat for gloss finishes should be opaque and hard, and excellent results are obtained by adding about 5 per cent of the finishing paint to it. Lead-free undercoats are better for lead-free finishes. One coat of gloss should not be applied over another without the first coat being flatted down.

Synthetic paints are similar to hard gloss paints except in composition. They have good gloss, durability and resistance to chemicals; and, not being so heavily pigmented, give a better coverage per litre, approximately 16-17 m^2. Drying times vary from 4-8 hours, after which they harden off rapidly. These paints are made with solvents which are more searching than turpentine or white spirit, and it is recommended that the manufacturer's special undercoatings should be used. Synthetic paints should be applied quickly, with less crossing and recrossing than with ordinary paints. Some manufacturers advise the use of two coats of synthetic finish instead of the usual undercoat and finish, but this is not general practice.

It is advisable not to mix different types of gloss paints together, as this may cause loss of gloss or failure to dry. Gloss paints tend to run or tear and it is essential they are brushed out well.

To convert m^2/litre into sq. yds/gal., multiply by 5. This gives an approximate conversion.

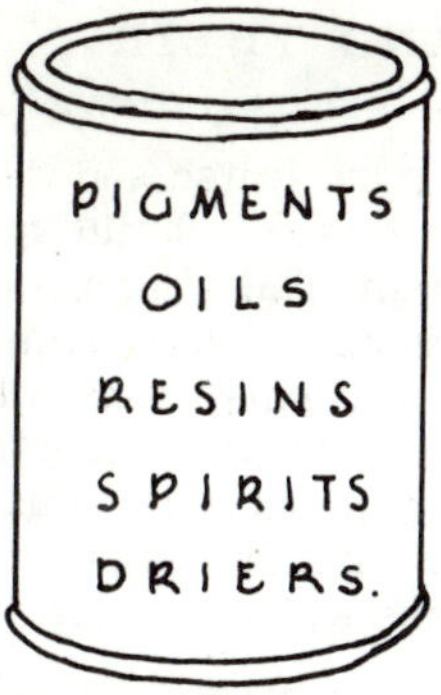

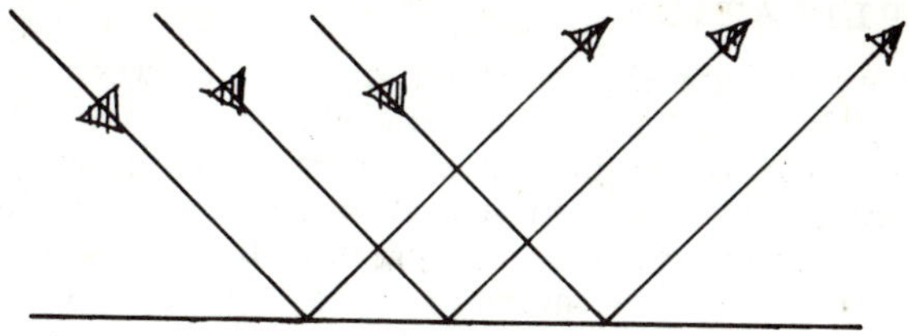

Gloss paints reflect light rays at the same angle but in the opposite direction.

(1) BRUSH ON PAINT

(2) CROSS

(3) LAY OFF VERTICALLY

The paint should be brushed on as evenly as possible otherwise runs may occur.

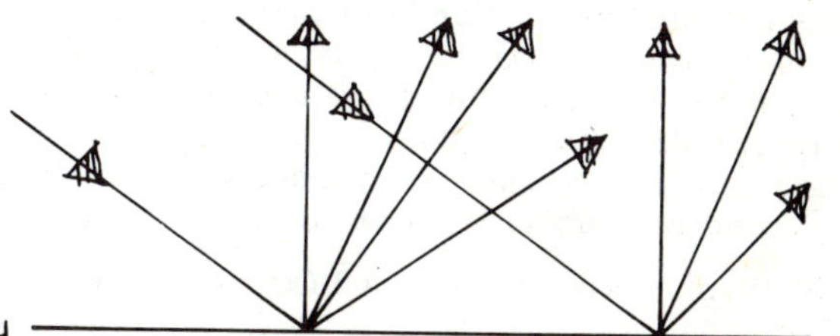

Semi gloss paints reflect light rays in a similar manner but scatter them.

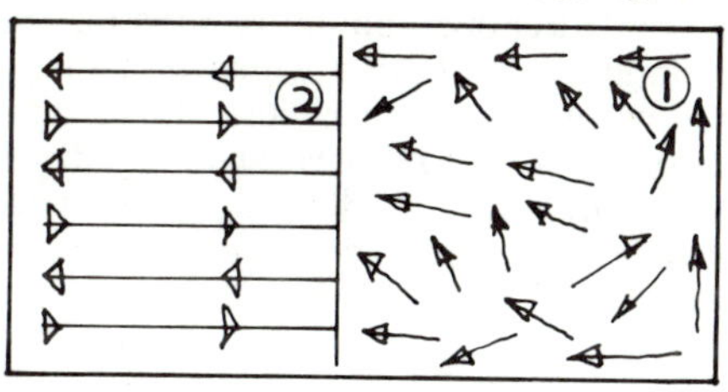

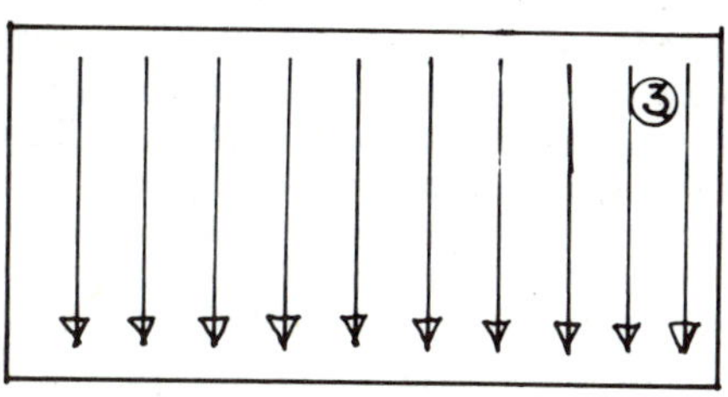

Over brushing is unnecessary as the material flows out to a level surface.

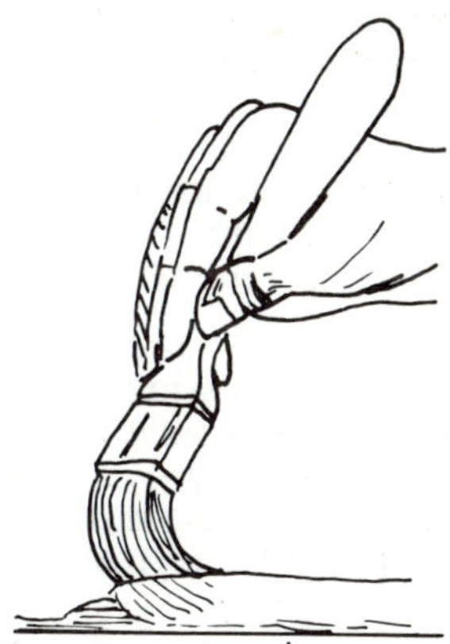

For gloss paints and enamels the brushwork should be firm and the bristles should travel through the paint.

GLOSS PAINTS AND FINISHES.

UNDERCOATS AND FLAT OIL FINISHES AND THEIR APPLICATION

Undercoats are the intermediate coats between priming and finish and they should dry hard, smooth, and have good opacity. They are used to build up the thickness of the paint film and form the base for the finish. Quite a number of undercoats are made up on the job from paste white pigments—white lead, zinc white, titanium white or lithopone, with oil, white spirit or varnish, and in this manner they can be mixed to suit all types of surface and degrees of suction. The oil content should not be too great, otherwise the undercoat will remain soft and eventually cause the harder finishing coat to crack. The addition of a small quantity of hard varnish will help the 'flow' and assist hardening.

Ready mixed undercoats are primarily intended for interior work, unless the manufacturer states otherwise. Many types can be adjusted to suit individual requirements, but the additions should not be too great, otherwise it may 'pick up' when painted over. For intermediate coats a little oil can be added; and, for undercoatings for gloss paints, a little white spirit. Water affects many undercoatings while in fluid form and all brushes, utensils and surfaces should be perfectly dry before being used. Most ready mixed undercoats are quick drying and should be applied, crossed and laid off quickly to give a smooth, even coat. Undercoats usually cover 18-20 m^2/litre.

Flat oil finishes are mainly used for decorative purposes, since they are of pleasing appearance, reflect light well, but do not have the same resistance to hard wear and the weather as gloss paints. Flat finishes are useful over uneven surfaces, as they tend to hide rather than emphasize any unevenness. They should not be overbrushed, and a full generous coat should be laid on in all directions. Overbrushing or overlapping with a spray gun on a rapidly drying edge may result in unevenness or 'flashing'—shiny patches on the otherwise flat surface. If the material is too sharp, a little oil may be added, but this must be done sparingly, otherwise patchiness will result. Coverage for flat oil paints is generally in the region of 16-17 m^2/litre and they dry in 6-8 hours. The best results are obtained by using an oily undercoat and stippling the flat paint, or, on plain surfaces, by the use of a paint roller. Flat oil paints can usefully be adjusted with white spirit, oil and varnish for undercoats.

Imitation stone paints are very similar in appearance to flat oil paints, but have coarse particles such as sand added, which produces a textured effect. The ground for this type of paint should be smooth and without suction. When applied, the paint must be stippled to even out the coarse particles. Too thick a coat should not be applied, otherwise flaking or peeling will result.

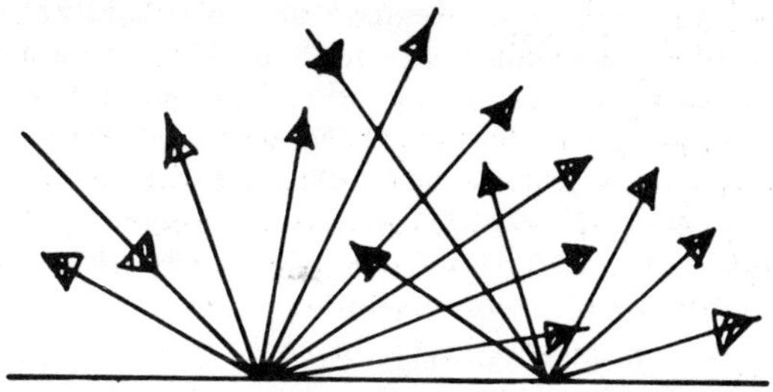

Flat paints reflect light rays in all directions.

BRUSH APPLICATION OF FLAT FINISHES.

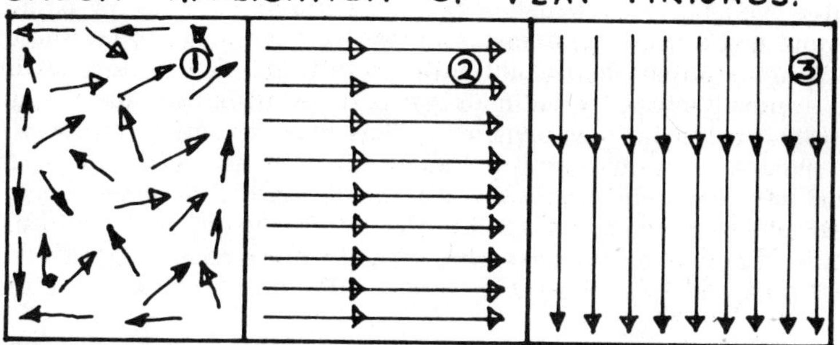

(1) Apply paint in all directions laying on paint liberally.

(2) Cross with horizontal brush strokes.

(3) Lay off with vertical brush strokes.

For a first class finish the wet paint can be stippled.
The stippler should strike the surface evenly, it only needs to travel about 80mm from the wall at each stroke.

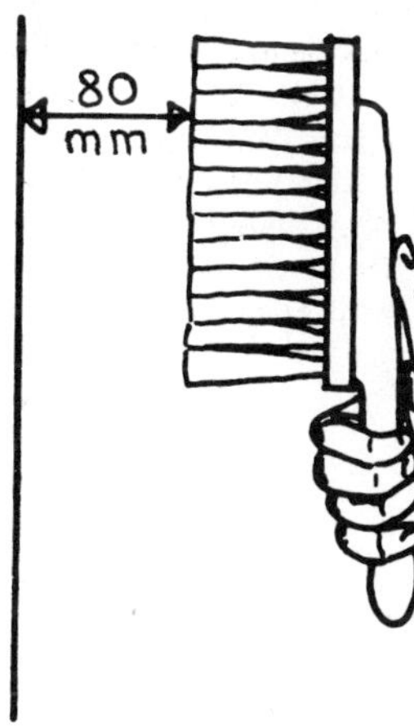

The stippler illustrated is one with an adjustable handle enabling even wearing of the bristles.

UNDERCOATS AND FLAT FINISHES.

The Construction and Care of Brushes

Although painting can be done by spray and roller, brush application still remains the most important method. Paint brushes are made of all bristle; hair and fibre; or nylon or other synthetic filaments. The bristles or filaments are cemented together with vulcanized rubber or synthetic resin, which enables their use in all types of materials and solvents, and are fixed on to the handle or stock with wire, sheet metal or seamless ferrules. Sheet metal ferrules are riveted on the handle but seamless ferrules are pressed on, engaging bristle and handle in one smooth band.

Bristles are obtained from the hog, pig or boar; their characteristic tapering and flagged end enables them to take up and hold paint well, and their springiness and hard wearing qualities make them still the best material for brush making. When ignited bristles burn with an offensive odour and completely disappear. Brushes of British manufacture made of all bristle always have 'pure bristle' stamped on the ferrule or handle. To enable large size brushes to be made at a cheaper price, bristles are blended with horsehair or fibres from the aloe tree, and such brushes are particularly suitable for painting stucco, cement and similar surfaces, as these surfaces tend to cut the bristles. Synthetic filaments do not hold paint quite so well as hog bristles, but improved techniques in manufacture, including flagging or splitting the tips, are gradually increasing the standard of performance. They are extremely hard wearing and very much cheaper than the bristle. New brushes do not need to be soaked before use and any hairs or filaments not cemented in will be shed during the first few strokes of working; but the handles of brushes which have been in use and laid aside for a time may have shrunk, and in this case soaking is necessary. Coating the handles with knotting makes them easier to keep clean and to handle.

Brushes should not be dipped in more than half their length of filling and should not be left standing in the pot at any time; and they should not be stood or left in water too long, since it swells and distorts the handles and rusts the ferrules. It is better to rinse them in white spirit or paraffin or a proprietary brush cleaner before cleaning them with soap and water; afterwards rinsing in cold water, tapping the filling on a firm sharp edge to remove all surplus moisture, and laying carefully aside to dry. Distemper and wall brushes should be cleaned with soap and water and hung up to dry. On no account should soda, strong alkali or acids be used for cleaning. If the material is difficult to remove, rinse the brush in paraffin before washing in soap and water. Brushes used continuously on mouldings or edge-wise part or 'finger' and work unevenly. Writers and lining fitches are best rinsed in white spirit and tallow or vaseline worked into the bristles to keep them in shape.

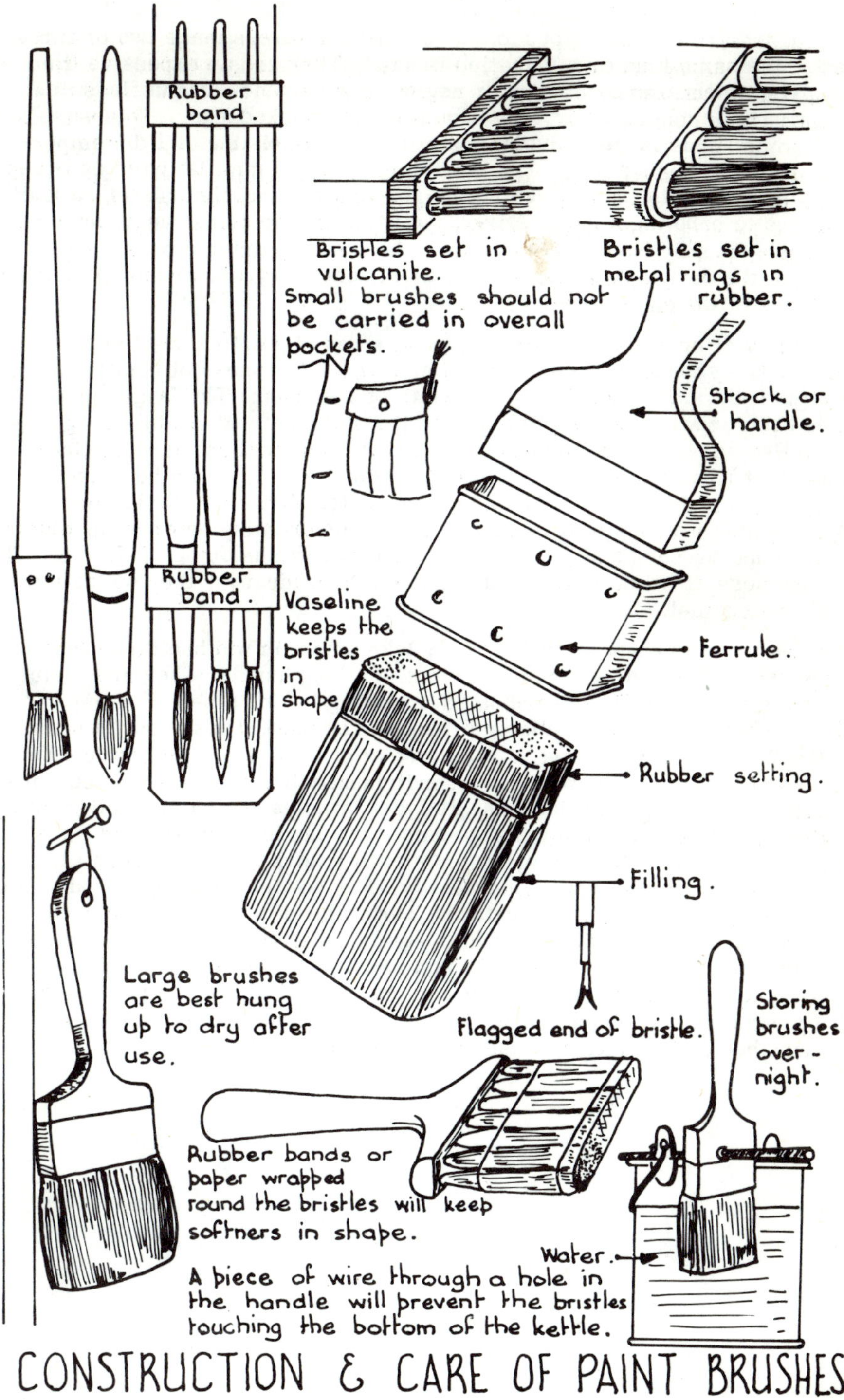

CONSTRUCTION & CARE OF PAINT BRUSHES.

Paint Brushes

Paint brushes are of many types and some manufacturers make two or three grades of the same kind to suit the job in hand. They are an expensive item of painting equipment, and by misuse or neglect can become quite useless in a very short time: consequently, they should be well looked after. To ensure this is done, some firms issue individual painting kits consisting of 1 distemper and 1 washdown brush, 2 sets of 75 mm, 50 mm and 25 mm flat painting brushes, and 3 paint kettles, which become the responsibility of the painter until the time comes to hand them back. Large contractors employing casual labour for painting jobs usually have a 'colourman', who issues paint and brush kits at the start of the day, replenishes kettles when necessary, and cleans all kettles and brushes at the end of the day.

The majority of brushes used for nearly all types of painting are flat paint or varnish brushes made in widths 12-100 mm with a thickness of bristle or filament from 6-21 mm, according to the width of the brush. The larger size brushes have bristles 62-75 mm long and the more refined grades are ground to a bevelled edge to save 'working in'. A new brush will not produce fine brushwork at first and for the best work a 'worn in' one should be chosen. If the fillings exceed 75 mm in length, they are fitted with a removable metal bridle. Varnish brushes oval in section are a compromise between the older type of ground brushes and flat brushes, the usual widths being 25-75 mm. The extra thickness of filling in the centre makes them ideal for the application of varnish and enamel.

Fitches are small brushes ranging from 2-28.5 mm in width, either round or flat in section, attached by a metal ferrule to a long handle. They are useful for painting mouldings, sash bars, and for getting behind pipes. For narrow surfaces difficult to reach, radiator brushes are made. These consist of the head of a 25 mm flat brush set at right angles on a pliable wire handle 400-600 mm long, which can be bent to suit individual jobs. For covering ceilings and other large areas with flat paint, 100 mm brushes are ideal. Wall painted with gloss paint require 75 mm; large sash bars, angle irons and steel joists can be tackled with a 50-75 mm, but for smaller work such as narrow glazing bars a 25 mm flat brush will be necessary. For covering corrugated asbestos roofs, felted roofs and similar surfaces, Turk's head brushes can be used. These are circular 100-125 mm in diameter similar to common grade distemper brushes, fixed on a wooden broom handle. Always use the largest brush that will do the job effectively and choose the best brush for the job. Good work requires good brushes, but if the job is a very rough one and will ruin a first class brush it is better to use one of inferior grade.

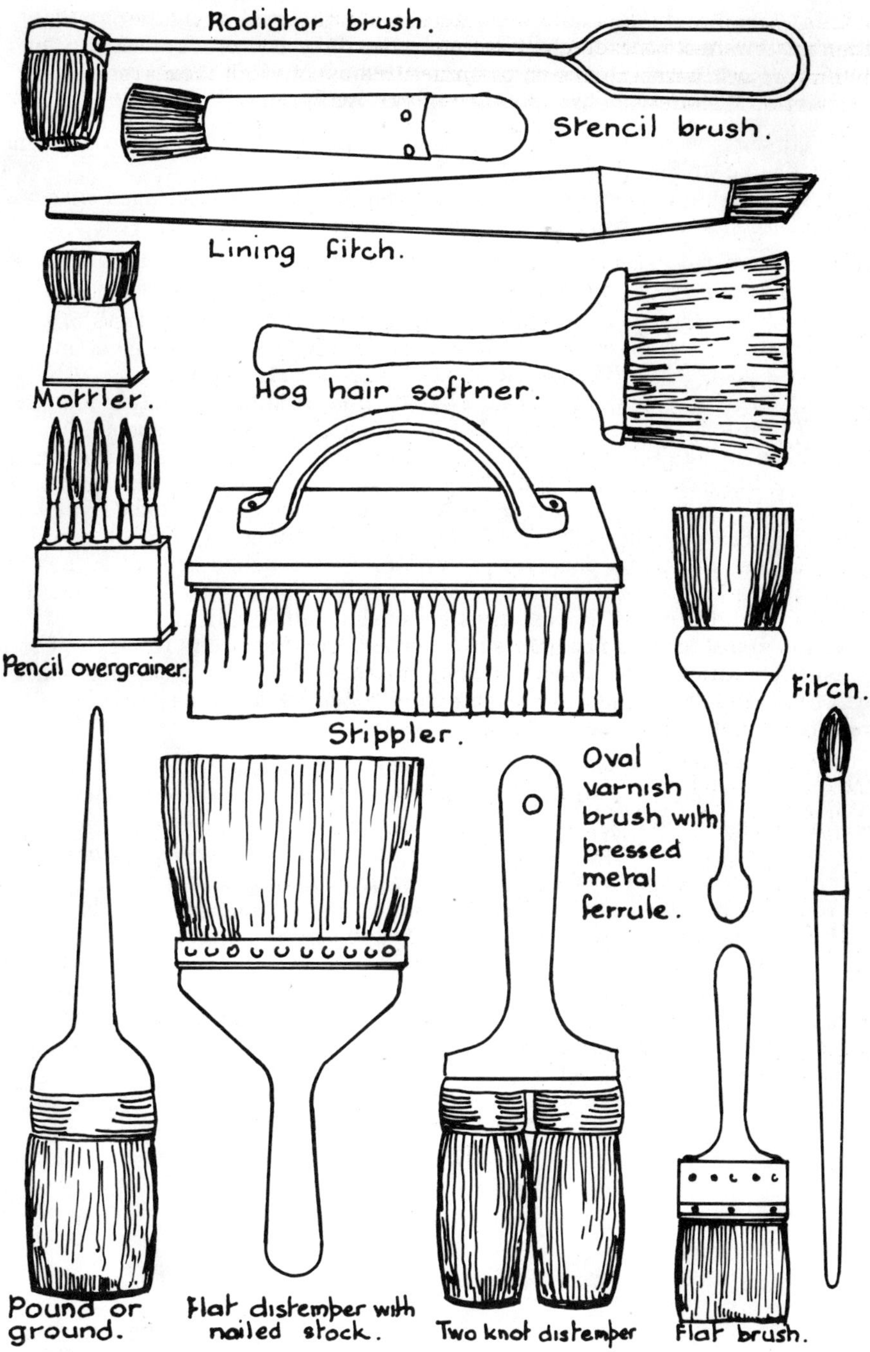

TYPES OF PAINT BRUSH.

Brushes for Covering Large Areas

For coating large plain surfaces with distemper, waterpaint, emulsion paint and other water mixed materials, brushes wider than 100 mm are used. These are distemper, wall, weatherboard or kalsomine brushes, of which there are many varieties. Distemper brushes are of two main types:

(1) Knotted, with two or three knots or bunches of bristle fastened to the handle with copper wire or metal ferrule
(2) Flat, with a continuous row of bristle about 25 mm thick fastened with a sheet metal ferrule or a band of hide fixed with clout nails, as with the Lancashire pattern.

Distemper brushes can be obtained 175-225 mm wide with a bristle length of 125-150 mm. The Lancashire type of distemper brush has three knots of bristle fastened to a splayed stock 225 mm wide. Others more convenient to handle have 2 knots, are 175 mm wide, with bristles 131-150 mm long. Flat brushes include distemper, which are 175-225 mm wide, with bristles approximately 137 mm long; kalsomine 125-200 mm wide, with bristles about 130 mm long; weatherboard 100-137 mm wide, with bristles about 130 mm in length; and wall brushes 175 mm long and 106-125 mm in length of bristle. The amount of bristle, mixture or nylon filament used in the manufacture of brushes varies, but a bristle brush marked 226 g will have taken that amount of bristle in its making allowing about $1\frac{1}{2}$ per cent for waste. Flat brushes are not so heavy to handle as the knotted types and are easier to use for cutting into a moulding or to a straight line. The best brushes have a fair length of bristle. If they are too short they do not hold sufficient material and tend to throw it from the brush. It is advisable to soak brushes that have not been in use for some time, to swell the stock and so re-tighten the fixing. All brushes should be thoroughly cleaned after use and hung up to dry.

Rollers are sometimes used instead of brushes. They consist of a metal cylinder 62 mm in diameter covered with a sleeve of lamb's wool, nylon, sponge rubber or mohair which can be slipped off for easy cleaning with soap or detergent and water. The handles vary: some have handles like a paperhanger's roller; with others the axle of the cylinder extends and is bent twice at right angles to form the shank for the handle. Sizes range from 125-175 mm for general work and 62 mm for small work. The material is held in a shallow tray with a ridged surface, or in a bucket with a ridged-board attachment, and the roller passed to and fro over the ridges to distribute the material evenly. Extra high surfaces can be reached with the roller fixed on a long broom handle or pole. For cleaning rollers, scrape the roller with the tool provided, rinse with an appropriate thinner, wash in soap or detergent* and water and hand up to dry. If the sleeve is removed for cleaning, it should be replaced immediately.

* Do not use for sponge rubber rollers.

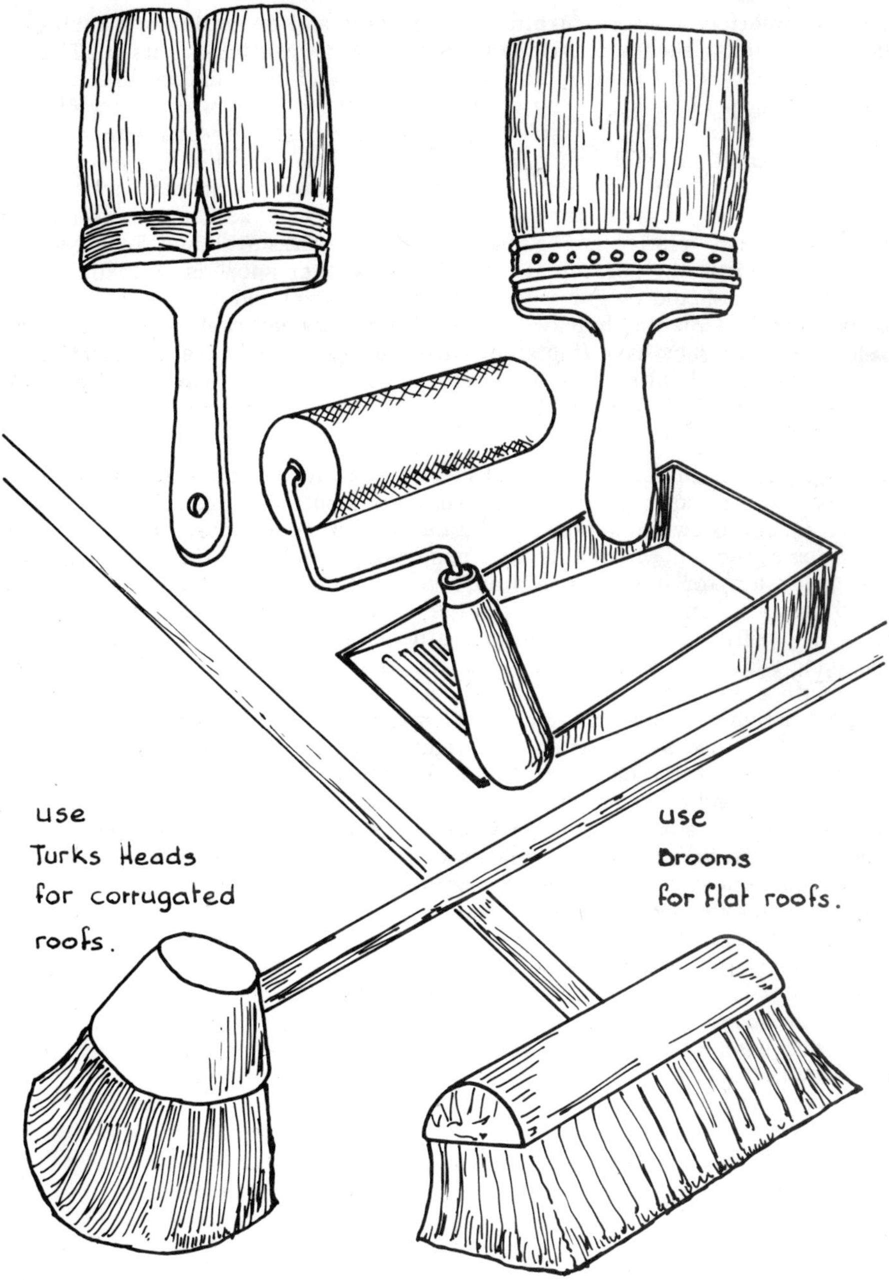

BRUSHES AND TOOLS FOR COVERING LARGE AREAS.

Tools for Preparation

SHEETS, WASH-OFF BRUSHES, WIRE BRUSHES, BLOWLAMPS AND GAS TORCHES

For the protection of floors, furniture, machinery and other fixtures during painting, twill dust sheets, plastic covers or light tarpaulins are used. They should not be allowed to accumulate dust or dirt, and need shaking frequently and laundering periodically. For washing off soft distemper, wetting wallpaper before removal and cleaning paintwork, wash-off brushes are used; these generally have 2 knots of bristle or hair set in cement and fixed on the stock or handle with copper wire. Sometimes distemper brushes past their prime are used for the same purpose. Strong alkali solutions such as caustic soda quickly destroy hairs and bristles and 'grass' or fibre brushes are used for these material. To rinse the surface afterwards, use either sponges or swabs. Buckets and kettles should be galvanized or pressed steel and of light weight for ease of handling. Flaking brickwork and stonework need rubbing with a wire brush. These brushes have fine steel wires instead of bristles and handles of various lengths and pattern. Mild corrosion on sheet metal can be rubbed with a wire brush, but badly corroded metal surfaces need chipping with stout tungsten carbide steel scrapers.

Burning off is the quickest method of removing cracked, blistered or overloaded paintwork and this can be achieved with a petrol or paraffin blowlamp, or a gas torch. Blowlamps consist of a cylindrical fuel container with a slight depression at the top, which acts as a reservoir to hold the small quantity of spirit necessary for lighting the lamp; a screw or escape valve for lowering or extinguishing the pressure; a safety valve with spring and washer for maintaining an even pressure, which, should it become too great, allows the gas to escape and reduces the pressure to normal again; and a horizontal or inclined burner with wind shield. Paraffin lamps also have a pump to stimulate the flow of oil, but petrol or benzolin lamps rely entirely on volatilization of the spirit. The general method of operation is to light the fuel in the reservoir and let the heat thus generated cause the fuel to vaporize, which, when the valve is opened, ignites and becomes the blowlamp flame. With paraffin lamps, pumping increases the pressure and generates more heat. All blowlamps need keeping clean and free from soot, and a small tool called a pricker can be bought for cleaning the nipple. Gas torches burn either butane or propane gas, which can be obtained in cylinders of various sizes. The smaller sizes can be screwed on to the torches but the larger ones are connected by means of a length of rubber tubing, some having two couplings enabling two operators to use the same supply. Gas torches burn with an even pressure and are almost trouble free. Of the many types of gas torches, some have interchangeable nozzles for broad work, sash work and brazing, and a small pilot flame when the trigger feed is shut off.

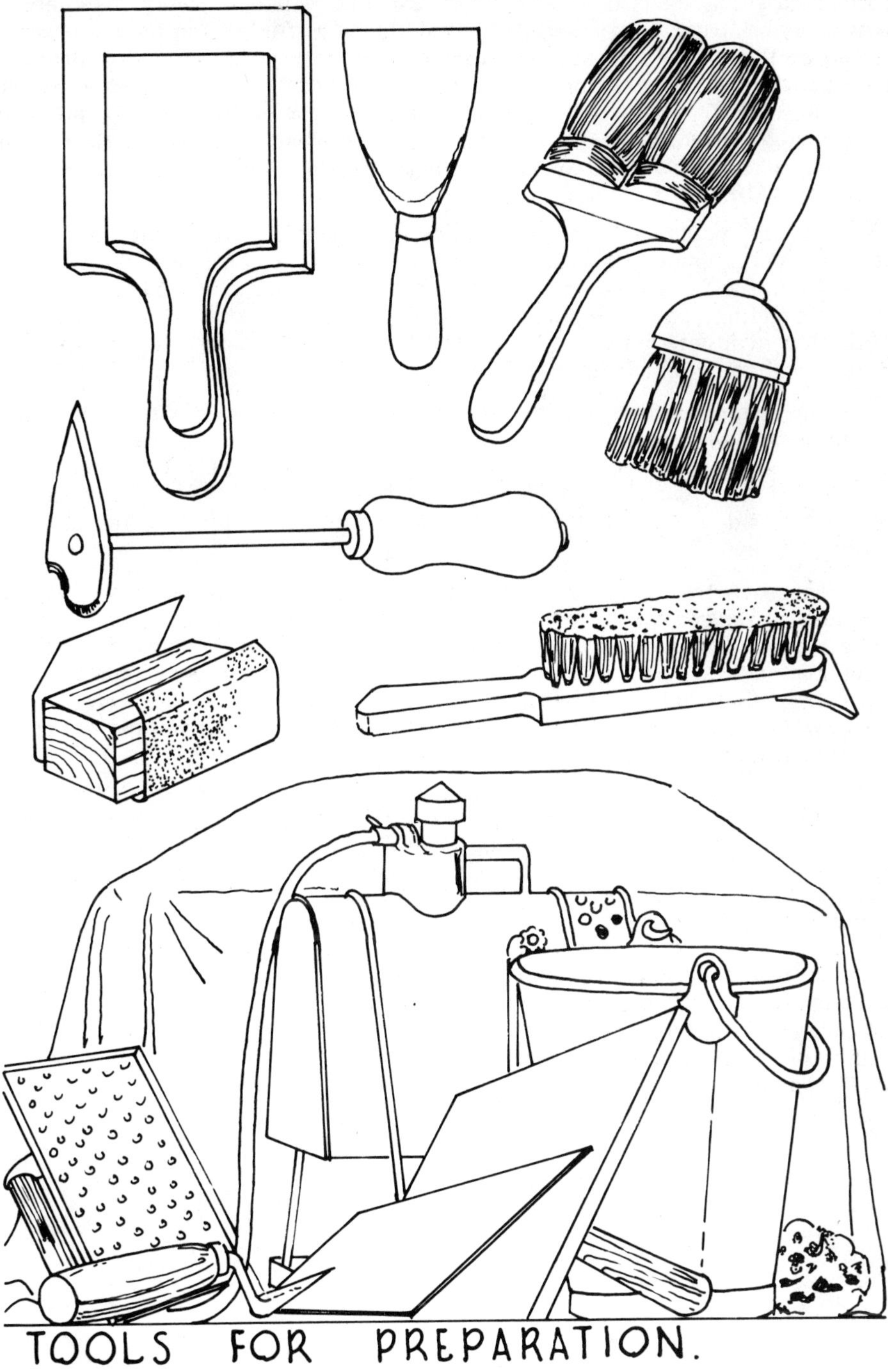

TOOLS FOR PREPARATION.

Tools for Preparation

THE PAINTER'S PERSONAL KIT

To make a surface suitable to receive paint, distemper or paper, it is necessary to make it perfectly clean, free from loose particles and injurious matter, and smooth; and this may involve the partial or complete removal of previous distempering and papering, the repair of defective plaster, the washing of paintwork, the removal of all flaking, scaling or blistered paint, and the replacement of broken glass and sash cords. Such operations require tools which, with the exception of large brushes, buckets and sponges or swabs, constitute the normal personal kit of a painter.

For removing paint from broad surfaces in conjunction with a blowlamp or paint solvent, loose or flaking materials or wallpaper, a stripper with a stiff blade 75 mm wide is used, and where the surface is narrow a 25 mm chisel knife. Mouldings and enrichments require the use of a shave-hook; some have fixed blades, either triangular or heart-shaped, others have interchangeable blades, which are an advantage.

For repairing plasterwork, a gauging trowel is useful; those with blades 120 or 150 mm long are a handy size. Where the trowel would be too big, a stopping knife can be used. These have a flexible pointed blade with one straight and one curved side. Glazing knives are a similar size but have a stiff blade with two curved sides, and are used for positioning and smoothing putties when glazing.

For filling defects in flat surfaces to obtain a perfect finish, knives of varying widths with very flexible blades are used to spread the filling-up compound; as the edges of blades are easily damaged, light wood or cardboard shields should be used to protect them when not in use. Broken glass is removed with a hacking knife, consisting of a stout blade with pieces of leather riveted on either side to form a handle, which can be bound with adhesive tape to prevent jarring from the hammer blows.

There are many types of dusting brushes, the cheapest being pitch set, which are only suitable for removing dust and debris before painting. Better varieties are vulcanite set or mounted in ebonite, which can also be used for brush graining or scumbling, as they will withstand the action of solvents and repeated washing. When rubbing down flat surfaces, a block of cork or soft wood 100 × 75 × 40 mm, around which glasspaper can be wrapped, is useful.

To hold filling and stopping materials, shaped boards, 200 × 250 mm, with handles 100 × 40 mm, will be required. For large repairs to plasterwork, a hawk, consisting of a platform 300 mm square of metal or wood with a handle on the underside, will be required to hold the compo or plaster.

A hammer, pincers, screwdriver, punch and a couple of chisels complete the kit, which can be carried in a wooden tool box or attache case.

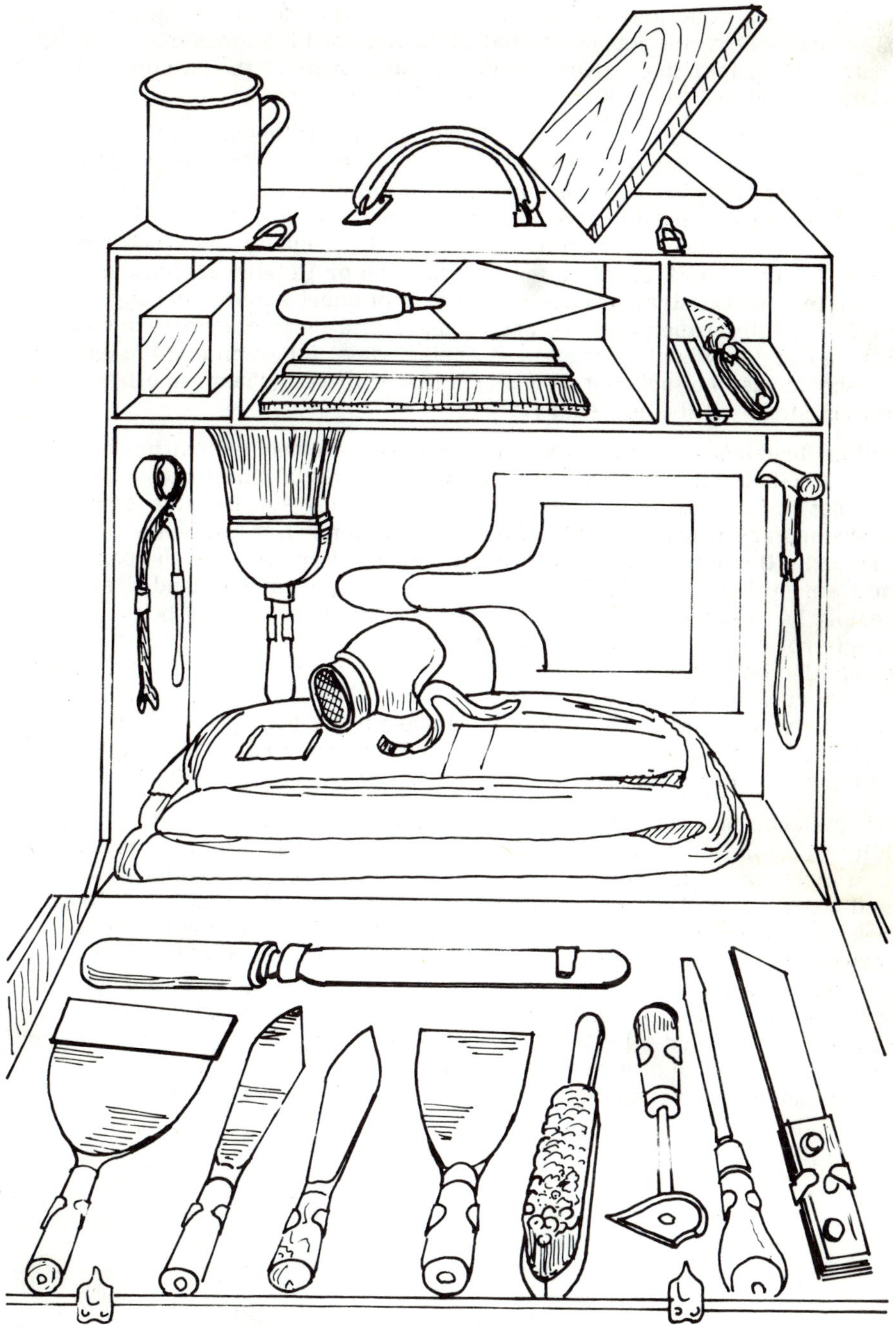

PAINTERS PERSONAL KIT.

Tools required for Paperhanging

Although the bench, buckets and pasting brush used by the paperhanger are supplied and sent on the job, the actual tools required for measuring, cutting and putting the paper in position are the responsibility of the painter and form an addition to his personal kit.

For measuring and setting out, a pencil, 1 m folding rule, chalkline, plumb bob, and spirit level are indispensable. Two pairs of shears are required, one pair with pointed blades 300 mm long for trimming or edging and a shorter pair with bull-nose ended blades around 250 mm long for cutting the paper when wet. Besides trimming with shears, speedier results can be obtained with small patent trimmers, of which the Morgan Lee or Ridgley are typical. A sharp knife, a metal straight edge and a strip of sheet zinc for cutting on are required for thick papers, or use can be made of a more elaborate device, such as a Ridgley Wheel Trimmer. A casing wheel for trimming the ends of paper should also be included. To keep knife and wheel sharp, an oil or carborundum stone will be necessary.

Smoothing brushes, for smoothing and pressing the paper in position, are flat brushes 175-300 mm wide, either in continuous rows of small knots pitch or rubber set or in a series of knots 25 mm in diameter set in metal ferrules. Two brushes are recommended, a thick full one for relief and heavy textured papers and a thinner one for light weight papers. For pressing the paper well down, a smoothing roller 225 mm wide felt covered is recommended, since it can easily be re-covered when dirty. For the joints and angles, boxwood seam and angle rollers 37-50 mm wide are made; a useful tool combining both rollers can also be obtained. An apron with pockets large enough to hold smoothing brush, rule and scissors is also necessary. The painter should also have in his tool kit a hammer and tacks for canvassing or fixing temporarily heavy relief materials. Push pins can also be used for heavy relief materials. Screwdriver, pincers, set square and pair of dividers are also useful.

Benches should be light but rigid and should combine maximum strength and rigidity with minimum weight. The simplest form consists of two trestles 0·75 m high and 0·5 m wide, placed 1·5 m apart, with either two loose or hinged boards 2 m long and 0·5 m wide, just stout enough to span the distance without sag, trestles with metal stays being better than those with cords. Paperhanging benches combining boards and trestles which fold up to form a case suitable for carrying rolls of paper, straight edges and paperhanging kit inside, are also obtainable and simplify handling. When paperhanging, cleanliness is essential and all equipment should be kept perfectly clean. Two or three buckets, a piece of hessian for straining the paste, and a sponge and chamois leather for removing paste from paintwork, will be required.

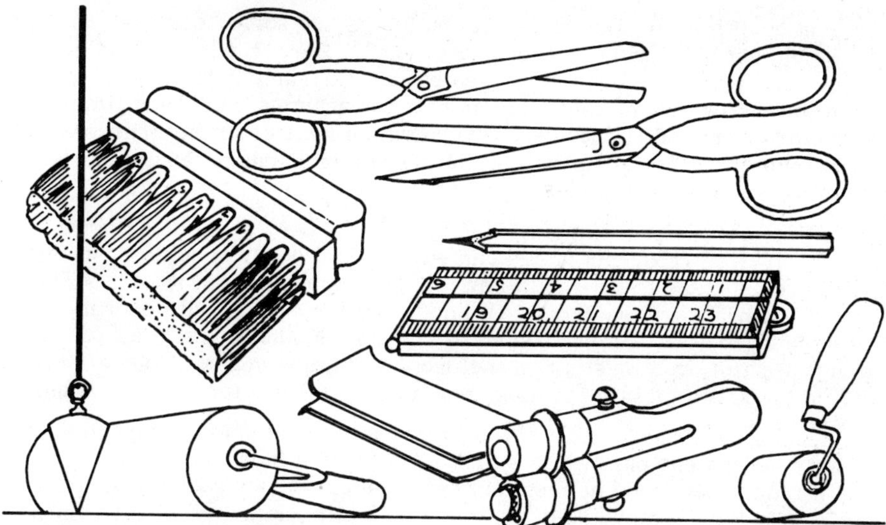

PLUMB BOB AND LINE, PAPERHANGERS' SMOOTHING BRUSH, BULLNOSED SCISSORS, POINTED SCISSORS, PENCIL, 1-METRE RULE, FELT-COVERED ROLLER, POCKET TRIMMER, ANGLE ROLLER.

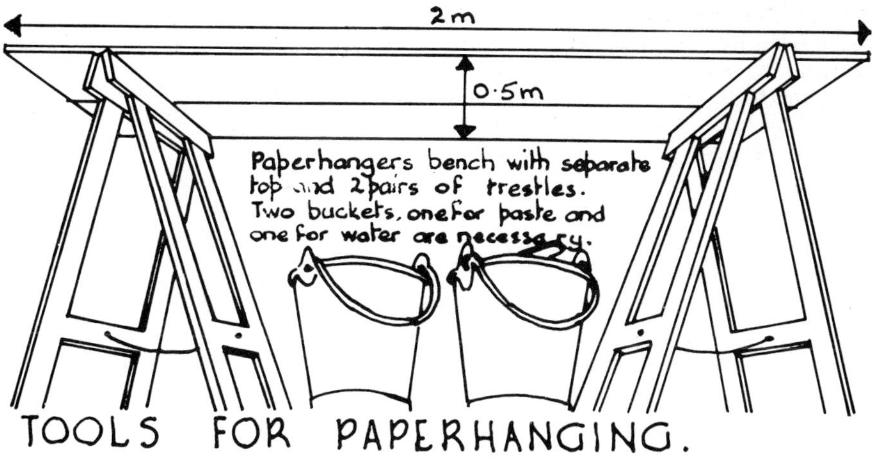

TOOLS FOR PAPERHANGING.

Preparing the Surface

NEW AND UNPAINTED WOODWORK

The largest quantities of timber used in buildings are soft woods deal, fir and spruce which contain a large percentage of resin. Hard woods, such as oak, mahogany, teak and walnut, are also used but not the same extent. Many buildings show attack by insects and fungi and as a precaution much new timber is treated before fixing with a preservative. Quite a number of these liquids can be painted over without losing their effectiveness or harming the paint.

New woodwork should be rubbed down with glue-bound glasspaper Nos. 1½ or F2, going first across and finishing along the grain, care being taken not to remove the arrises from mouldings and edges. Nail heads should be punched well below the surface to provide a key for the stopping, which should be done after the priming is dry and hard. All knots require coating with knotting, two thin coats being better than one thick coat. After the knotting, a thin coat of sharp drying paint over these areas before the priming coat will obviate blistering around the knots, as oily priming does not key satisfactorily to knotting.

Soft woods can be primed with a red lead and white lead primer thinned with three parts oil and one part white spirit. The British Research Station advises the use of aluminium wood primer on very resinous woods such as Douglas Fir or Oregon Pine if they are dry. The self-sealing nature of aluminium wood primers keeps the resin well isolated, but also imprisons moisture contained in the wood, which can interfere with adhesion of the paint and cause peeling of the paint system.

Hard woods can be primed with red lead and white lead primer thinned with one part oil and three parts white spirit; or with aluminium wood primer. Another favoured method is to coat the surface with a mixture of equal parts gold size and white spirit. This pre-treatment is followed by the priming coat in the usual way. Teak is of an oily nature and is generally oiled or varnished, but if it has to be painted a wash over with acetone or solvent naphtha will remove much of the oil from the surface. It is best primed with aluminium wood primer.

To prevent blistering and flaking due to the seeping of moisture into the timber, it is essential that all back edges and cut ends should be painted before fixing. For this reason it is advisable to twice coat all end grains with priming. Timber not previously treated which has badly weathered and now has to be painted should be coated with boiled linseed oil, left a few minutes for the oil to sink in and wiped dry before priming is applied. The priming should be oily and well brushed in. Creosoted timber which is to be painted should be treated with knotting and followed with a fairly sharp primer or coated with aluminium wood primer. Emulsion paints have also been recommended as 'buffer' coats for such surfaces.

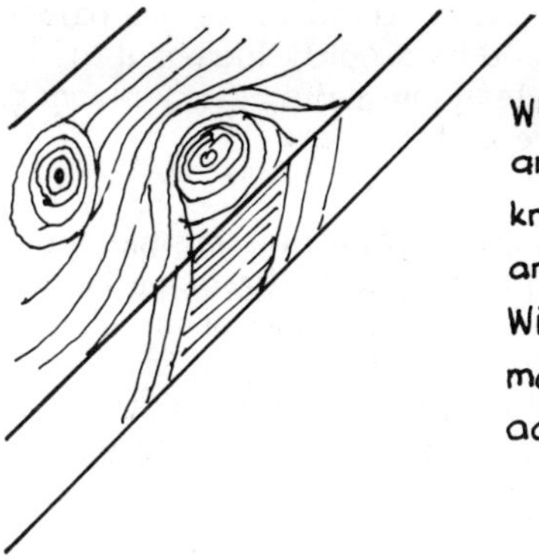

When using primers such as red and white leads always coat the knots with knotting – two thin coats are better than one thick one. With aluminium wood primers this may not be necessary, but it is advisable.

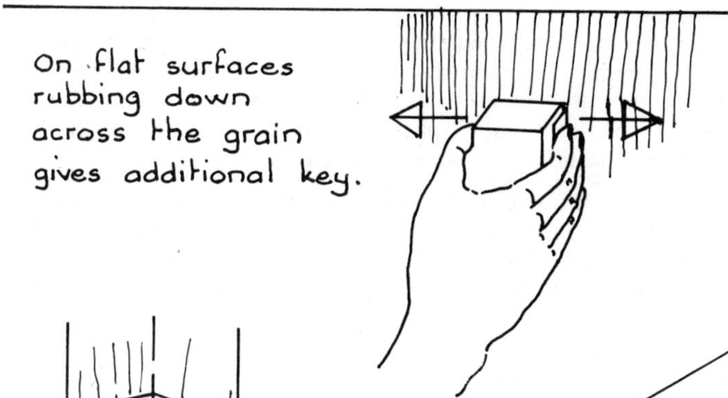

On flat surfaces rubbing down across the grain gives additional key.

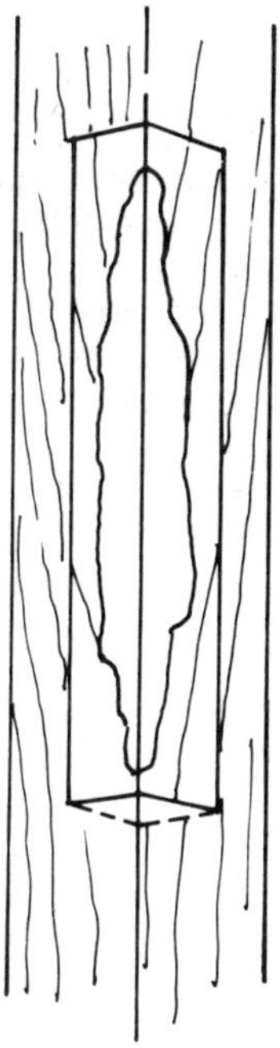

All repairs should be made before any painting is done.

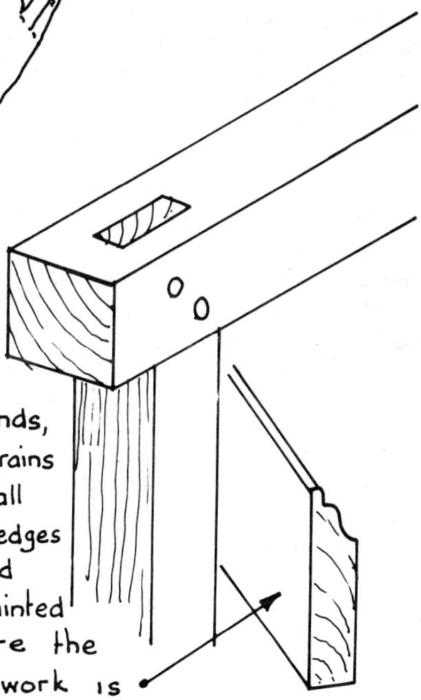

Cut ends, end grains and all back edges should be painted before the woodwork is fixed.

PREPARING THE SURFACE – NEW & UNPAINTED WOOD.

Painting Windows and Doors

Where possible, all door and window furniture should be removed before painting; but where this cannot be done the screws should be slightly loosened to enable the painter to paint underneath them in a clean and satisfactory manner. Methods of painting double-hung sash windows are as follows.

Outside painting
(a) Pull down top sash and push up bottom sash, (b) paint meeting bar and top parts of bottom sash, (c) reverse position of sashes by lowering bottom sash and raising top sash, leaving the bottom sash well clear of the sill and runners, (d) paint top of sill, outer runners and 80 mm up of inner runners, finishing in a semi-circle, (e) paint bottom edge and remainder of bottom sash, (f) paint bottom edge of top sash, (g) paint top sash, (h) paint frame, front of sill and any mouldings to frame.

Inside painting
(a) Raise the bottom sash as high as possible and lower the top sash, (b) paint meeting bar and lower parts of top sash, (c) reverse position of sashes by raising top sash and lowering bottom sash, leaving the top sash well clear of the frame and runners, (d) paint top of frame, inner runners and 80 mm down outer runner, finishing in a semi-circle, (e) paint top edge and remainder of top sash, (f) paint top edge of bottom sash, (g) paint bottom sash, (h) paint frame and architrave.

When doing the runners, the sash cord should be pulled clear and held while painting to prevent paint getting on the cord. Casement windows are painted glazing bars first, then rails, stiles and edges. Usually these windows open outwards and it is usual to paint the hanging edge with the inside and the top closing and bottom edges with the outside. With centre hung pivoted windows all edges opening into the room would be painted in with the inside and those opening out with the outside. With panelled doors the mouldings should be done first, then the panels. This should be followed by the muntins, rails, stiles, and finally edge and rebate of frame. Another method is to paint the panels first, then the mouldings, doing the edge of the door before the stiles. Glazed doors should have the glazing bars cut in first, then any panels, followed by rails, stiles and edges. With enamel, gloss paints and varnish, it is usual to bring rails, muntins and stiles along together to prevent the edges drying and catching up. Paintwork on the doors and windows should be laid off along the direction of the grain, i.e. following the construction of the joinery. When outside and inside work is being done it is usual, weather permitting, to finish the outside work first, gutters, weatherboarding and soffits, stackpipes, windows, doors, railings and other iron work. For inside work, windows, picture rail, doors, mantelpiece and skirting. This ensures brushes do not pick up dust during painting.

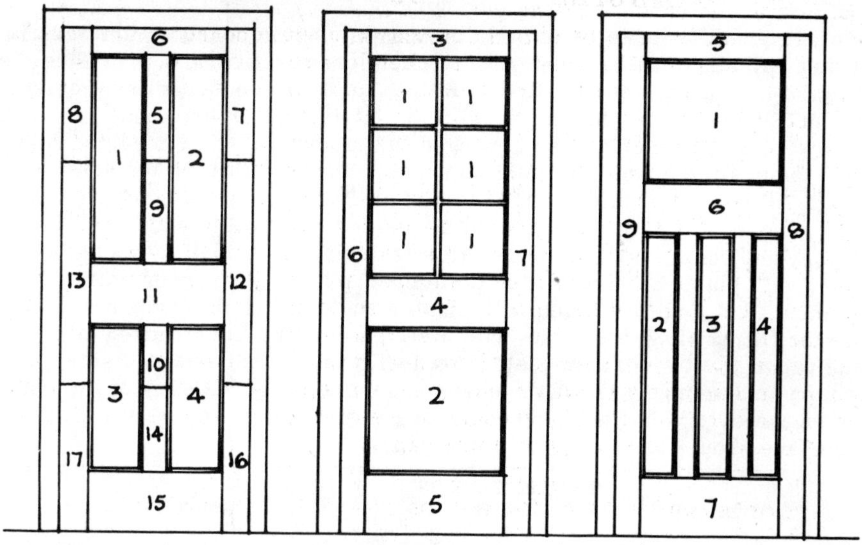

METHODS OF PAINTING DOORS.

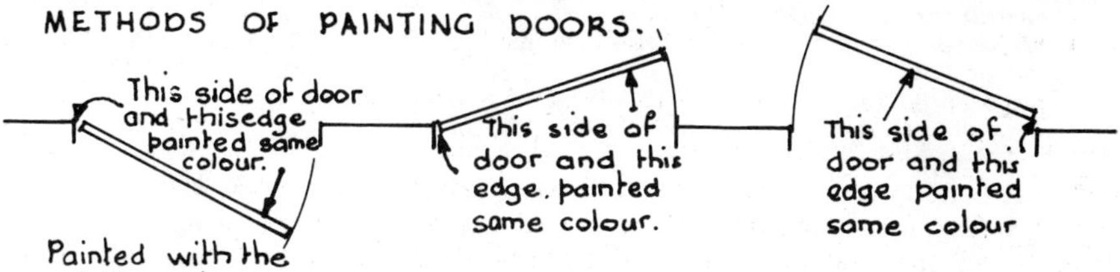

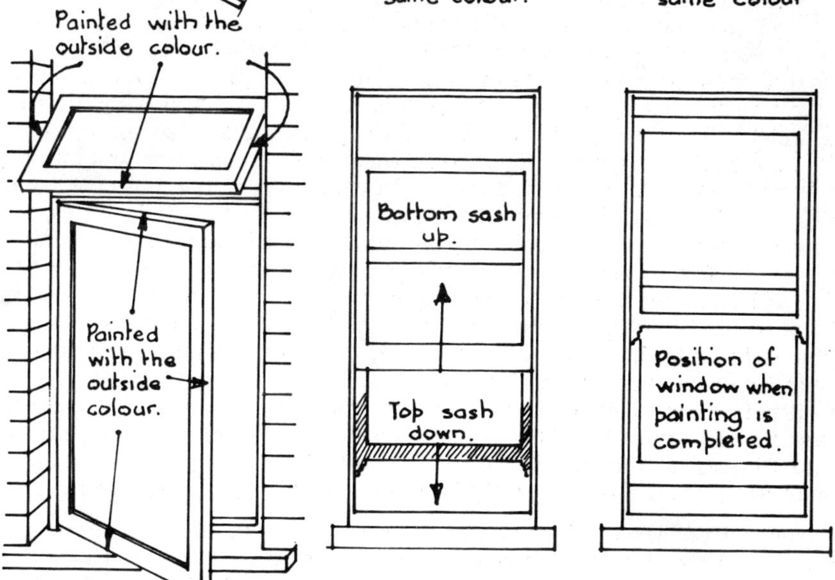

PAINTING WINDOWS & DOORS.

BUILDING BOARDS

One of the great advantages of building boards used in construction work is that they can be painted without risk of chemical reaction, and, providing the suction of the surface is satisfied, painting can be proceeded without delay. The chief difficulty lies in dealing with the joints; and, unless these are covered with a wooden batten or are so arranged that a chamfered or moulded edge masks them, after a time the joints open and cracks following the construction appear. All boards should be fixed to a stout frame or backing which will not warp or move. All nail heads should be painted, preferably with white lead, gold size and white spirit, and when this is dry all dents and joints can be stopped with plaster and washable distemper, plastic paint, proprietary filler, or painted first and then filled with plaster and paint. It is advisable to reinforce the filling along the joints with a strip of scrim, linen backed tape or lining paper, the two former pasted down with white lead and gold size paste. When dry and securely fixed, the edges should be filled with any of the materials mentioned, to hide the sharp edges of the strapping. For a first class job, the surface should be lined with lining paper.

TYPE OF BOARD	METRIC SIZES OF BUILDING BOARDS	IMPERIAL SIZES
Insulating Board	2440 × 1220 mm	8′, 9′, 10′ and 12′ × 4′
Hard Board	2438 × 1219-1220 mm	6′, 8′, 9′, 10′ and 12′ × 4′; 6′ × 2′; 7′ × 5′3″.
Super Hard Board	1830, 2440, 3050, 3660 × 1200 mm	8′ and 12′ × 4′
Decorated Hard Board	1830, 2400, 3660 × 1200 mm	8′ and 9′ × 4′

Very soft boards should be given a coat of fairly strong glue size (1 kg to 10 litres of water) or a proprietary wall board sealer prior to painting. The primer should be an oily one, consisting of 3 parts oil and 1 part white spirit, with sufficient undercoating to colour it. Medium hard boards can be primed with undercoating thinned with 2 parts oil and 1 part white spirit and super hard boards with undercoating mixed with equal parts of oil and white spirit. Chip boards, which are extremely hard but with a fibrous surface, should be coated with a mixture of equal parts of synthetic varnish and turpentine and afterwards smoothed by filling the surface with a casein type distemper. When dry, the filling should be rubbed down with F2 glass paper and sealed with well thinned gloss paint or a mixture of 1 part varnish and 1 part turpentine. Some manufacturers advise wetting the backs of the boards before they are fixed, and if this is done painting should be delayed until the boards are dry. A good plan is to paint or seal the backs to prevent moisture seeping in. It is sometimes desirable to treat building boards and timber used for exhibition and stand work with fire retarding material, to lessen fire risk. Treatment with two coats of flat oil paint or oil bound water paint will usually satisfy the requirements for Class III and often Class II. Two coats of silicate paint give a satisfactory finish for Class II and sometimes Class I, and two coats of paint based on calcium sulphate plaster will generally satisfy Class II and Class I requirements.

Polystyrene 'tiles' are often used as a covering for ceilings. They are usually 305 × 305 × 10 mm, but other sizes are also made, and can be cut easily with a sharp knife and straight edge. Adhesive is applied to the back—a dab at each corner and one in the centre—and the tile stuck in position. Careful setting out is essential before the tiles are fixed. If further decoration is required, water mixed materials should be used as oil paints tend to disintegrate them.

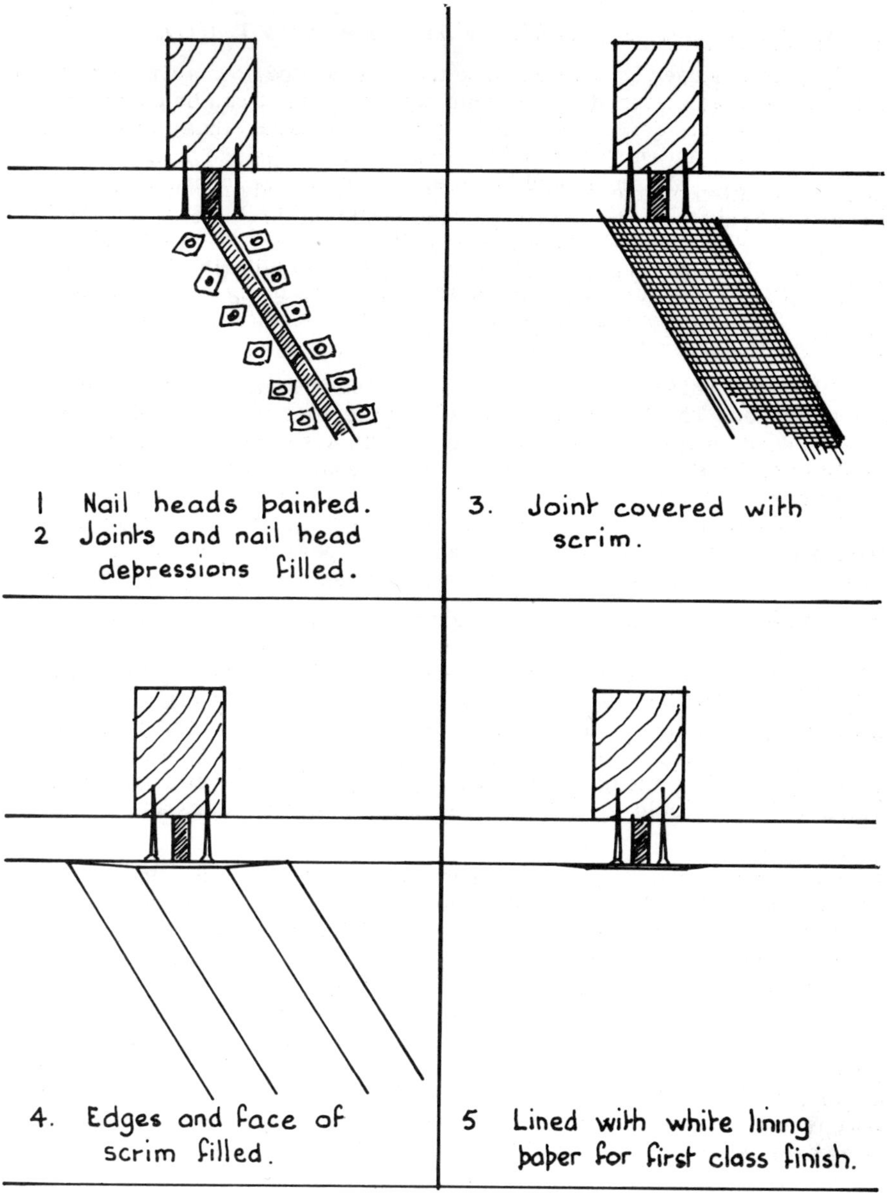

TWO METHODS OF DEALING WITH JOINTS OF BUILDING BOARDS.

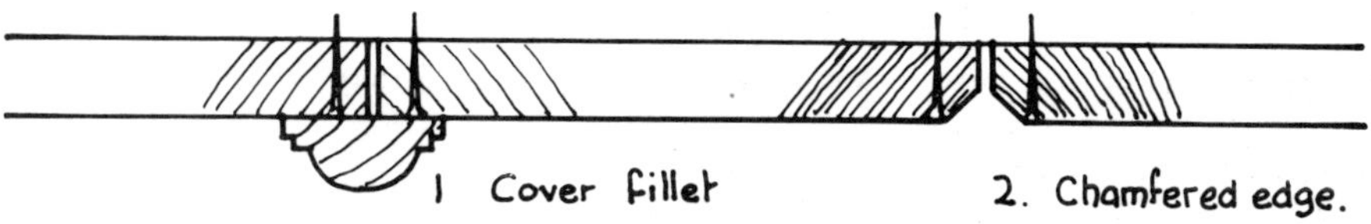

BUILDING BOARDS.

Preparing the Surface

TREATMENTS AND PRIMERS FOR NEW SOFT PLASTERS

Most of the plasters used in building have to be decorated at one time or another and if the plaster has matured and become perfectly dry it can be coated without any serious defects developing; but frequently the painter has to paint plaster which is new and laden with moisture in circumstances which are not very suitable for painting. Moisture tends to keep active alkalis from the backing, which combine with certain substances in the plaster to form caustic solutions which attack the paint, causing softening, discolouration and blistering. Moisture trapped behind the paint system may even cause the plaster to crumble. Ceilings and walls constructed in such a way that voids or spaces occur behind them which will allow moisture to escape freely are less likely to be affected than those constructed with solid backing.

Acoustic plaster should never be painted, otherwise its sound absorbing qualities are destroyed. The best treatments are single coatings of soft distemper, or water paint or emulsion paint, well thinned with water, spray applied.
Lime plaster or lime putty is used as a finishing coat on walls and ceilings, but as it only sets in very thin skins the surface is always soft and weak. Damp lime plaster should never be painted; otherwise saponification may take place. Soft distempers, water paints, emulsions and wallpapers are suitable first treatments. Dry lime plaster is very porous and requires a very oily primer.

Plaster, Plaster of Paris and Gauging Plaster are very quick setting plasters used for running mouldings, casting fibre slabs and decorative plaster and as a gauging plaster with lime or lime and hair. Unless these surfaces are perfectly dry, they should be coated with soft distemper, washable distemper or emulsion paint; otherwise such surfaces are very porous and need primers with plenty of oil. Stand oil or linseed oil in the proportion of 2 parts oil to 1 part white spirit as the media mixed with white lead would be effective, or unpigmented boiled oil could be used. Proprietary plaster primers are also suitable.

Wall plasters, sold under a variety of proprietary names, are similar to gauging plaster but have materials added which slow their setting. They are classified as retarded hemihydrate plasters. They work smoothly but the finished surface is apt to vary in suction. They are used for ceilings and walls and for making plaster slabs. Painting treatment is the same as for gauging plaster.
If these plasters are painted too soon, trapped moisture from the backing may cause 'sweat out' in which the plaster fails to harden and crumbles away. All new plaster surfaces should be lightly rubbed down with glue bound glass paper to remove nibs and coarse particles before painting. As moisture and efflorescence have a detrimental effect on paint, the painting of hemihydrate plasters should be delayed as long as possible.

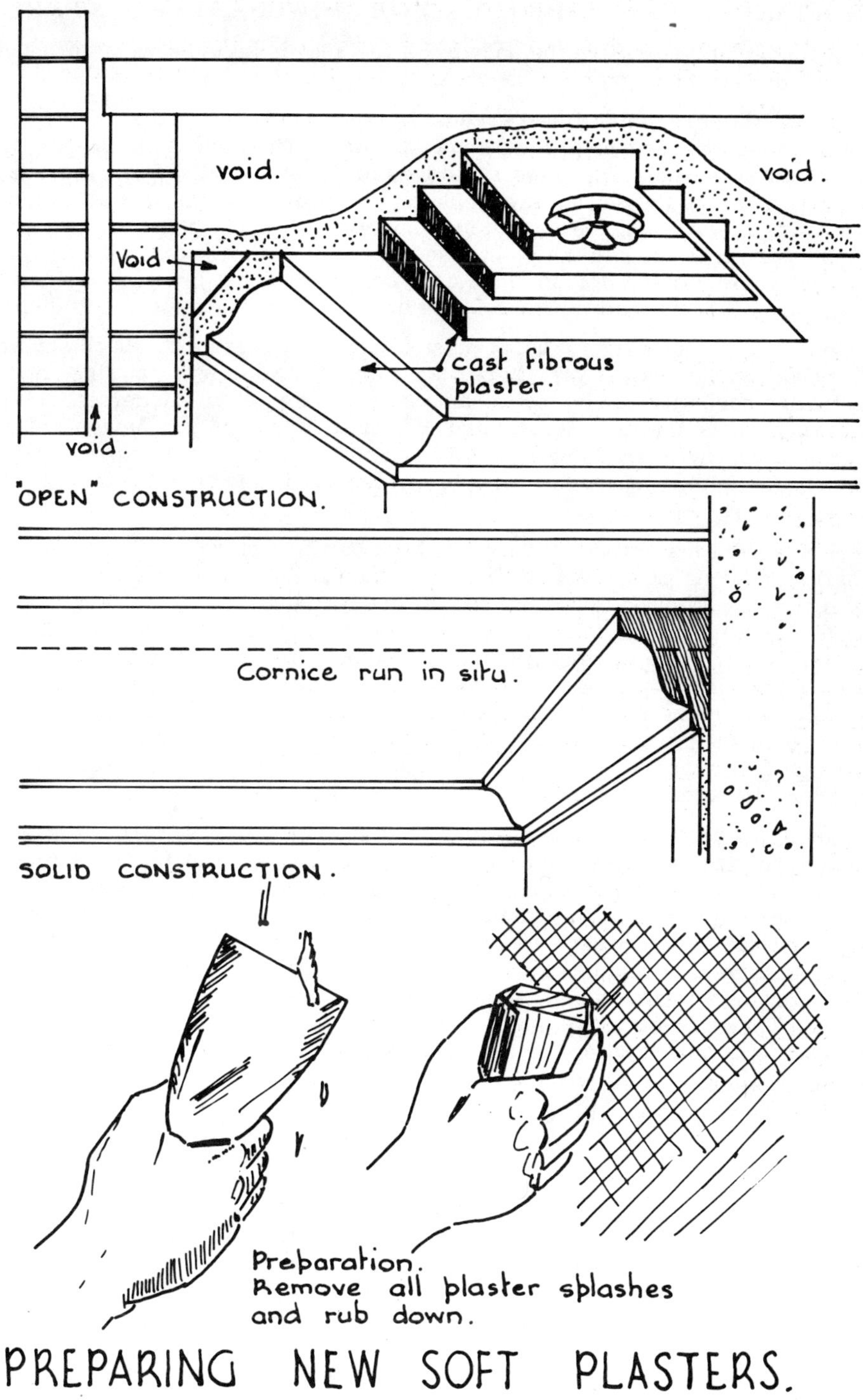

PREPARING NEW SOFT PLASTERS.

Preparing the Surface

TREATMENTS AND PRIMERS FOR NEW HARD PLASTERS

The main hard plasters are the anhydrous and anhydrite types, which produce surfaces of varying degrees of hardness.

Anhydrous plasters are of three grades: lightly, moderately and hard burnt, and all are subject to a technical fault known as 'dry out', which is the powdering of the finished surface due to too rapid loss of moisture caused by evaporation and suction before the setting is complete. To counteract this, a coat of thin paint, made up with white lead, gold size and white spirit applied 'following the trowel', that is, as soon as the plaster is set enough to withstand the action of the brush without disturbing the face, is recommended; but it must be done within 12 hours of the trowelling and left 3-4 days to allow the moisture to disperse.

'Sirapite' plaster is a lightly burnt plaster which produces extra hard surfaces and, providing the work is dry, there should be no trouble with painting. A thin penetrative coat with more media than pigment is required. Moderately burnt plasters, such as 'Kelite', 'Statite' and 'Victorite' are used for the general run of house property where hardness and fineness of surface is not so important. They require a thin penetrative primer, preferably based on white lead, brush applied.

Keene's cement and Parian cement are hard burnt plasters which set extremely hard, producing glass-like surfaces which do not hold paint very well, especially if they are allowed to become dry before being coated. For new work, a coat of thin, sharp paint, following the trowel, is the best treatment if a paint finish is required. A cheaper method, and sometimes as effective, is to coat the surface with water as soon as the setting stage is reached. If a water mixed material is to be used as a finish, a wash of water prior to its application will lessen the tendency to flake. For dry surfaces, the surface can be broken with a wash of 1 part spirits-of-salt and 4 parts water, after which the surface should be left to dry thoroughly.

Anhydrite plasters such as 'Pioneer', which produce hard but porous surfaces, are not prone to the defect 'dry out' in the same way as anhydrous plasters, but have a tendency to encourage 'efflorescence'. Oily primers based on China wood oil or similar to those used for soft plasters are suitable for this type of plaster. Should efflorescence in the form of soft, fluffy white crystals occur on the surface, it should be brushed off and the surface left two or three days. and the process repeated until all such deposits cease before any paint is applied. These plasters are often used for finishing coats on plaster boards and other 'dry construction' techniques, and in these circumstances difficulties should not arise.

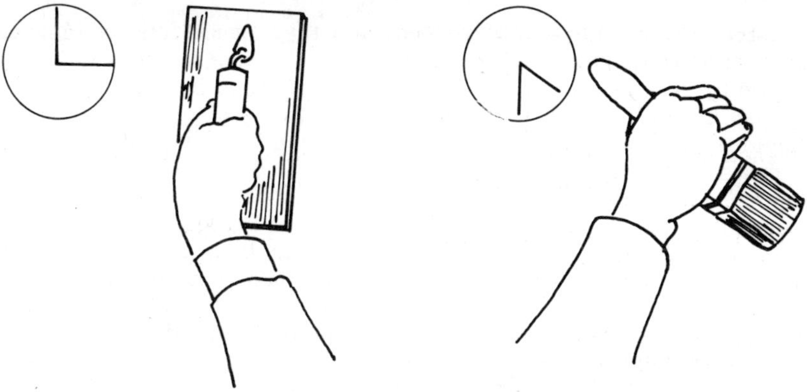

For new hard plasters "following the trowel" with a coat of thin white lead paint (paste white lead, gold size & white spirit) prevents "dryout" of the plaster and keys the surface for other coats. No more painting should be done for 3 or 4 days to allow moisture to disperse.

NEW PLASTER.

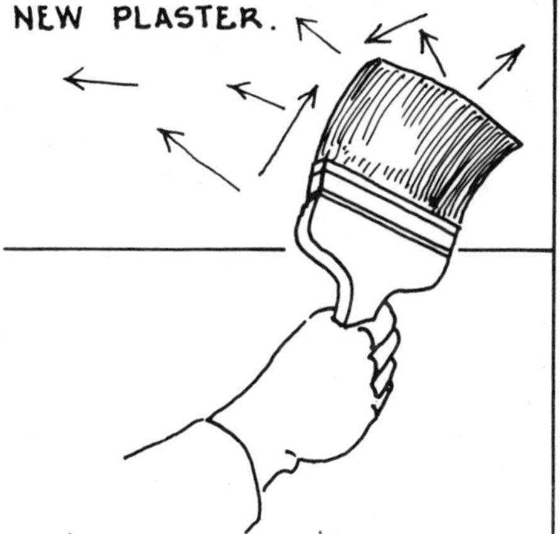

Instead of painting, brushing over the newly trowelled surface with water will break the glass-like finish of the plaster.

MATURE PLASTER. Rubbing down and applying a solution of 10 parts water and 1 part hydrochloric acid will break the surface.

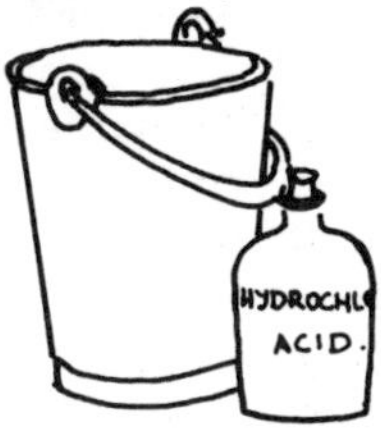

Mould growths should be brushed off and the surface coated with Santobrite or Shirlan N.A to kill the spores.

PREPARING HARD PLASTER.

Preparing the Surface

TREATMENTS FOR MATURE AND DAMAGED PLASTER

When plaster surfaces have to be redecorated, they should first be cleaned thoroughly and repaired where necessary. Soft distemper and wallpaper can be removed by repeated soakings with water, scrubbing the distemper off with a brush and stripping the softened paper with a stripping knife. The surface should then be washed with clean water to remove any traces of distemper, size or paste left behind. Washable waterpaints or distempers are difficult to remove and, unless the surface is overloaded, it is better to rub it down with glue bound glass paper and apply a thin penetrative but oily sealer to bind the surface. If the waterpaint has to be removed, it can be coated with a slack mixture of flour and caustic soda and left overnight, then removed by soaking and scraping in the usual way. Another method is the use of a 'Lightening' wallpaper stripping machine, which concentrates steam in small areas and so softens the media of the distemper.

With the surface cleaned, all cracks should be opened out with a knife or shavehook and undercut to give a wedge-shaped section which will hold the stopping firmly. Cracks should be thoroughly wetted and then filled with plaster or plastic material. The stopping should be pressed well home to obtain effective grip on both sides. Where one side of a crack has dropped, or is proud of the other, it is better to remove a wider area of plaster, say 100-150 mm, and make good level to both faces. Larger areas involving the backing or scratch coat should be well wetted and repaired with a mixture of plaster and washed sand to about 3 mm below the level of the surface and well scratched to provide key for the finishing plaster. When set the patch can be wetted and levelled off with neat plaster. Before making good, the edges should be painted to prevent the surrounding plaster being stained by the water, allowing the plaster to set more evenly and to prevent it breaking away at the edges. When doing repairs, always use the same type of plaster as the original work. This will ensure that both materials are of similar strength; hard plasters do not key satisfactorily to soft plasters. Badly cracked plaster surfaces need lining with lining paper.

Plaster which is stained or has mould growths needs special treatment. If the stains are small, coating with fine plaster and water or white knotting will suffice; and, if large, coating with flat paint or white lining paper. Mould growths should be brushed off and the surface coated with a fungicide such as 'Shirlan N.A.' or 'Santobrite' and allowed to dry. For severe attack, two applications will be necessary. Gaps between the tops of skirtings, under picture rails and around architraves should be carefully filled. Wetting the gap and applying a proprietary filler, or painting the gap and applying a stout mixture of paint and plaster, are both suitable methods.

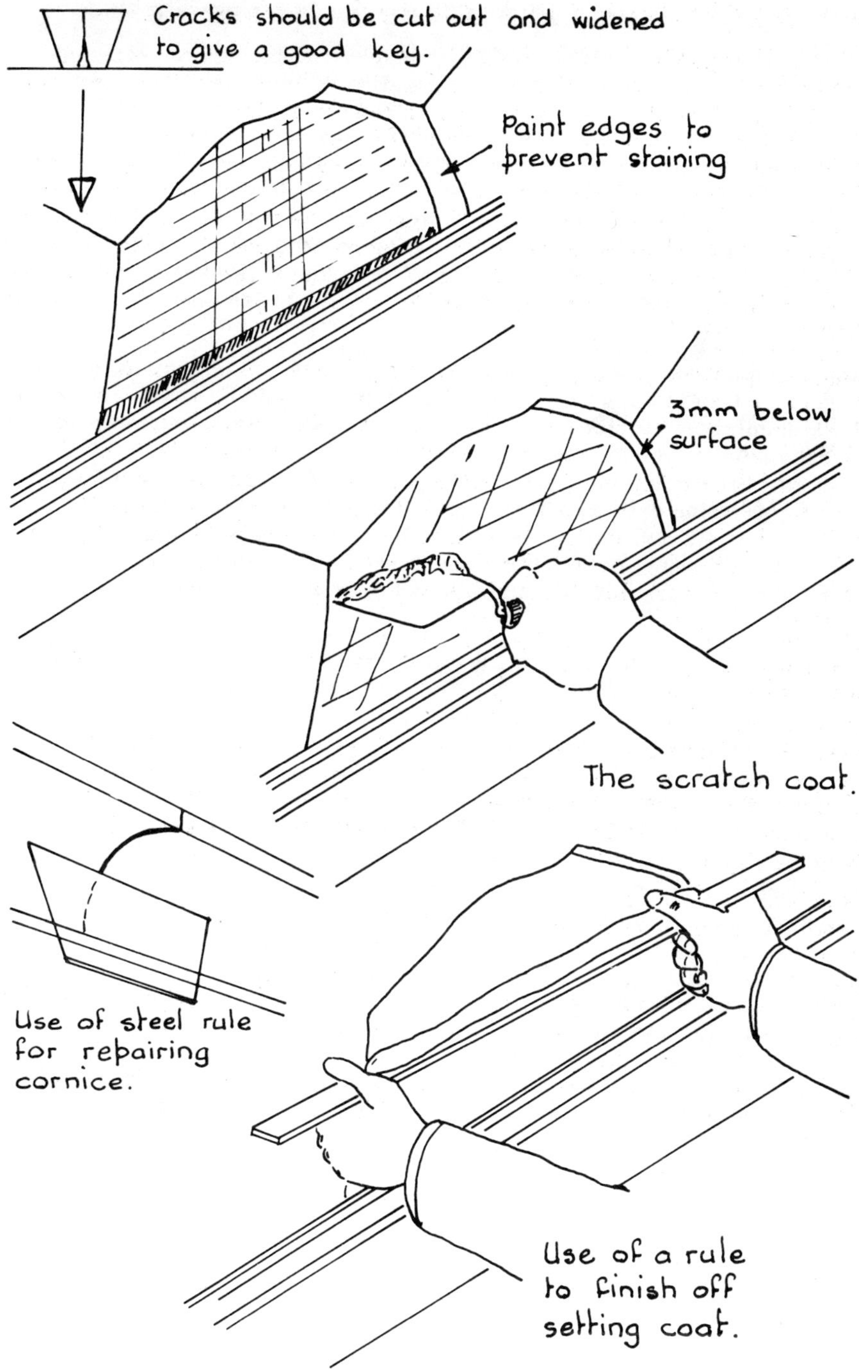

REPAIRING DAMAGED PLASTER.

PORTLAND CEMENTS AND ASBESTOS SHEETING

Portland cement is used in the jointing of brickwork; facings for external work such as rough cast and pebble dash; and plain and textured cement rendering. It is also used in bulk of various thicknesses for structural work. Smooth surfaces finished with a steel trowel do not hold paint so well as surfaces finished with a wooden float, which produce a coarser texture; and the primings used on such surfaces should be thinner and more penetrating. Cement and concrete surfaces having a high lime content should if possible be left at least 12 months to mature before being painted and a red lead and boiled linseed oil primer can then be used. Mould growths on the surface should be treated with a fungicide and brushed off with a wire brush before any paint is applied.

When new cement surfaces have to be treated, paints and materials that are alkali resistant and which do not saponify readily should be used. For inside work, water mixed materials like soft distemper, waterpaint, cement and silica paints can be used. Besides being porous and allowing moisture to escape, they all, with the exception of soft distemper and silica paints, form suitable grounds for painting over at a later date. For outside work, cement and water-paints, silica paints and bitumen emulsions are recommended, but although the first two can be painted over later as required, bitumen emulsion 'bleeds' through and discolours any oil paint applied over it unless the bitumen is sealed first.

Paints recommended for new cement are chlorinated rubber paints, paints based on epoxy resins and alkali resistant primers. Aluminous cements such as 'Cement Fondu' are not so alkaline as Portland cements and the risks with painting are not so great. With the painting of new plaster and cement surfaces, the greatest factor is the moisture content. If the fabric is dry and likely to remain so, then the painting can be done without any risk, but if wet then definite risks must be taken. Recently a new series of paints have been formulated which, it is claimed, will resist the forces of moisture movement and even allow such paint to be applied under water.

Asbestos sheeting, pipes and gutters need treating in the same manner as cement surfaces. Sheets treated only one side let in moisture which keeps the alkalis alive and also tends to push off the paint, and it is better to treat both sides and all edges before fixing. An alkali resistant cement primer or a synthetic varnish containing phenolic resin suitably thinned could be used for the purpose. Roman cement is mainly used for repairing stucco. It can be painted without any special treatment, a suitable primer being red and white lead thinned with 2 parts oil and 1 part white spirit.

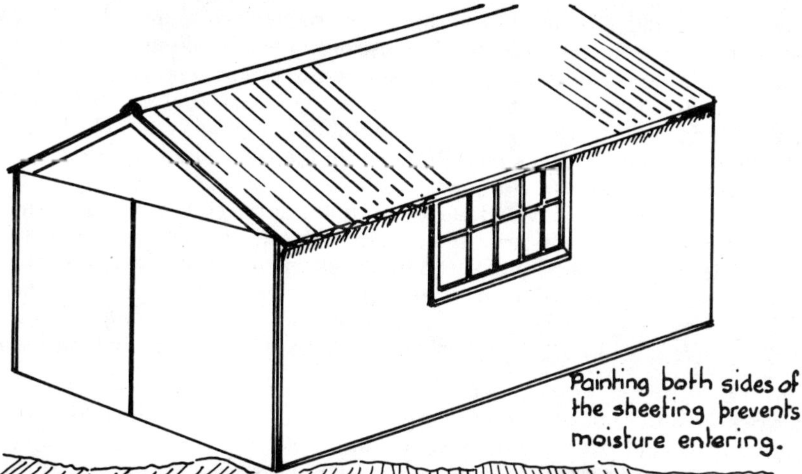

Chlorinated rubber paints are useful for swimming pools.

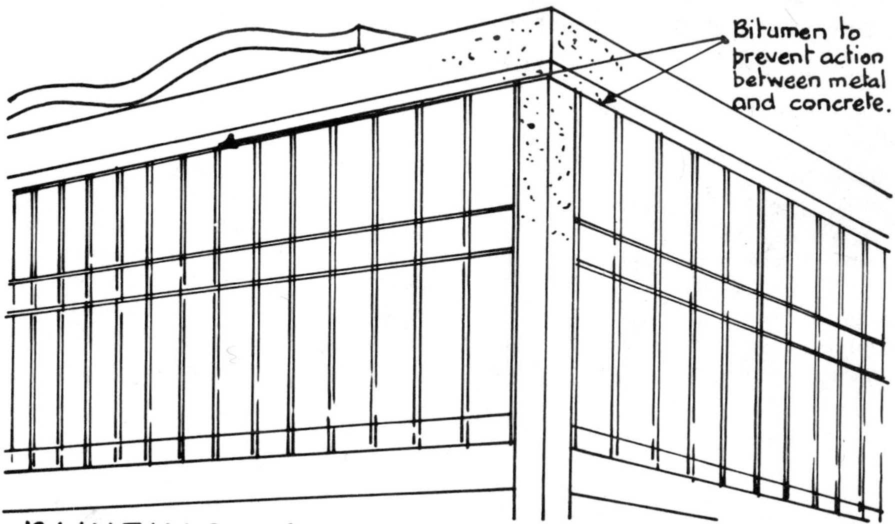

PAINTING ASBESTOS AND CONCRETE.

Preparing the Surface

BRICKWORK AND MASONRY

Brickwork is usually painted to reflect light or to give a special colour effect. The problems of painting brickwork are exactly the same as of painting cement, plus the additional problem of uneven suction over the whole area due to the difference in texture between the mortar and brick. Generally bricks can absorb a considerable quantity of water and if the surface is at all damp painting should be delayed as long as possible. Some types of brick, especially flettons, tend to throw off paint and can prove troublesome; and the addition of a little alkali-resistant synthetic varnish or powdered mica to the primer tends to overcome this. Soft bricks require a primer having a high percentage of alkali-resistant oil. Hard faced bricks require less oil but more white spirit. Two coats of primer are recommended as a precaution against misses. Old brickwork should be brushed down with a bass broom; all defective joints should be raked out; and, after being made good, the surface left to dry before painting commences.

Stonework should not require painting, but when it needs waterproofing, or has been repaired badly, is discoloured, or its colour does not fit in with adjacent surfaces, painting is usually resorted to. The usual method is to paint it to look as stone-like as possible and frequently stone paints which produce a very fine textured finish are used. The stone should be brushed down with stiff brushes or cleaned with a portable electric sander or wire cup brush, all traces of efflorescence brushed away, and any repairs done with a mixture of sand and cement mixed in proportions which give a fair reproduction of the texture of the stone. As the stone has to be wetted before it is repaired, it should be allowed to dry thoroughly before any paint is applied. Soft porous stones require a fairly oily primer similar to that recommended for soft brickwork, and hard stonework a primer similar to that recommended for hard bricks, or an alkali-resistant cement primer. If the stonework has been treated before, all loose or flaking material should be scraped off, the surface rubbed down dry, and an oily penetrative sealer applied.

STONE-LIKE PAINTS can be made from

Oily undercoating, Brick grit, Varnish, White spirit

Oily undercoating, Silver sand, Varnish

Water paint, Silver sand, Petrifying liquid.

Brickwork and masonry are sometimes treated with colourless damp-proof solutions to prevent rain driving into the fabric of the building; and all such coatings should be applied generously and in dry weather. At least 2 coats are necessary, with a period for hardening of the film in between coats. Damp proofers should not be applied during frosty weather. With some materials a toxic vapour is given off and if this type of material is sprayed the lowest pressure should be used to avoid mist. If the work is in a confined space and the vapour is inhaled over a long period, dizziness and sickness may result. This warning is usually given on the container.

Waterproofers protect the surface but do not alter the colour.

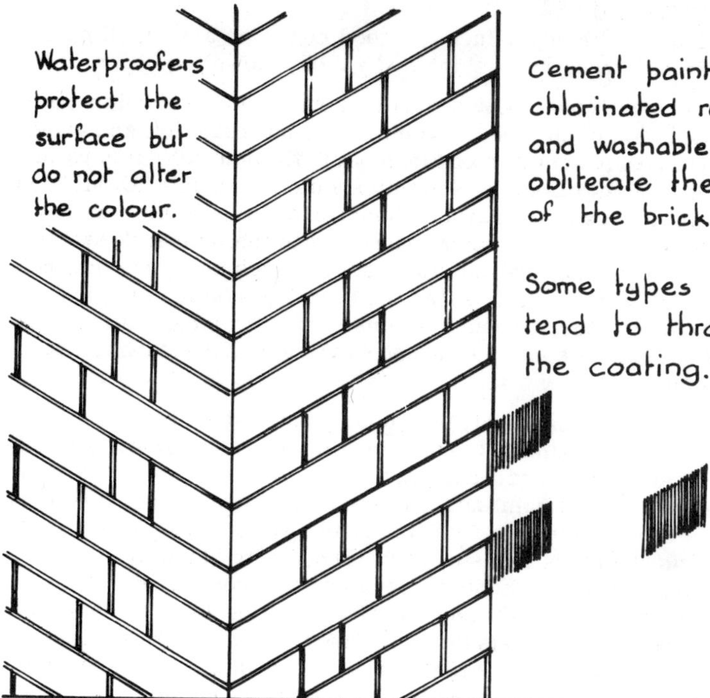

Cement paints, chlorinated rubber paints and washable distempers obliterate the texture of the bricks.

Some types of bricks tend to throw off the coating.

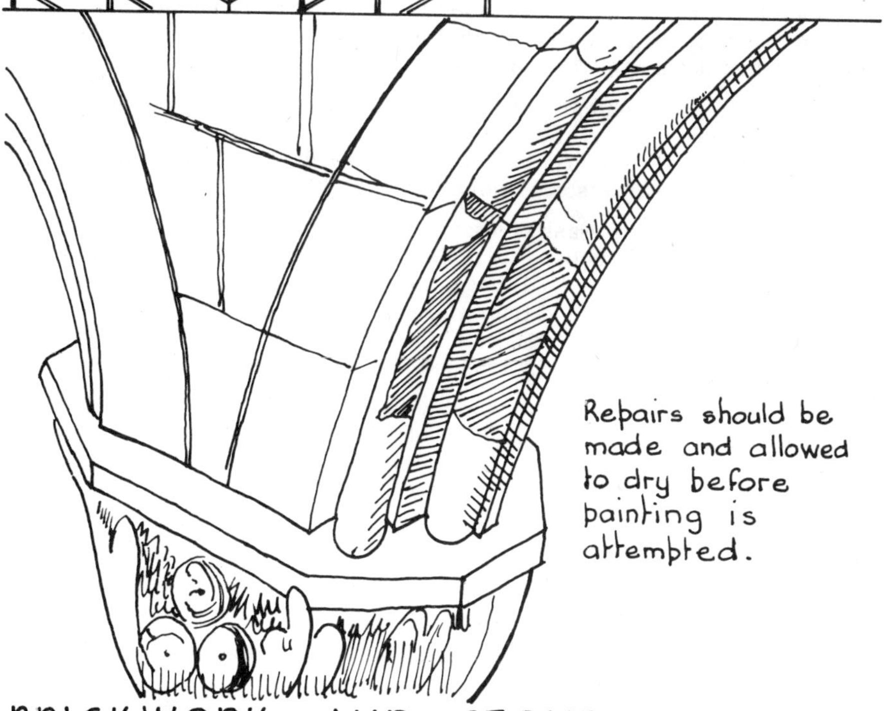

Repairs should be made and allowed to dry before painting is attempted.

BRICKWORK AND STONEWORK.

Preparing the Surface

PREVIOUSLY PAINTED WORK

In many ways, previously painted surfaces in good condition are the simplest to repaint, since they only require washing and rubbing down to prepare.

Washing should be done with sugar soap, soda or detergent dissolved in water, the strength of the solution being just sufficient to remove the dirt and grease but not the paint. Where surfaces are of gloss paint, flat varnish, or have been repeatedly treated with wax polish, the solution should be sufficiently strong to remove all the wax and just cut the surface. Always commence the washing at the bottom, using a wash-off or part worn distemper brush, and work systematically until the whole surface is covered. In this way, drips and tears will not course down the surface and mark it. After the cleaning solution or 'pickle' has been on for about 5-10 minutes, it should be rinsed off with clean water, starting from the top, the usual practice being to brush the water on and then finish off with a sponge or swab.

To obtain additional 'key' for the paint, any rubbing down should be done at this stage with pumice stone, pumice block or waterproof glasspaper, but great care is necessary to see that all grit and particles of abrasive matter are removed from the surface before painting. Pumice is a natural stone, very light and porous, which is sold in lumps and needs cutting with a hacksaw to face it up. Two pieces are used and the faces of each rubbed together to prevent clogging of the cutting faces. The lightest pieces cut the best. Pumice blocks are made of ground pumice and soda and if they are left in water too long they disintegrate rapidly. Waterproof glasspaper is used in exactly the same manner as glue-bound glasspaper but with water as a lubricant. It can be used over a cork block for flat surfaces or pressed to fit all types of mouldings. The Lead Paint Regulations 1927 require that all surfaces painted with lead paint should be rubbed down by a wet process.

Proprietary glazes and scumbles and flat varnishes, which are not very suitable for painting upon, sometimes prove difficult to prepare, and if washing does not remove the film it may be necessary to rub the surface with waterproof glasspaper or coarse steel wool and white spirit.

It may not be necessary to wash down a painted surface if it has started to 'chalk' and can easily be rubbed off. This is an indication that the medium of the paint has broken down, and all that is required is a dusting down and a very oily coat of paint applied to bind the surface. Outside woodwork from which the paint has peeled but which still retains remnants difficult to remove should not be wetted, as moisture may be trapped under the priming, but given a thorough rub down.

PREPARING PREVIOUSLY PAINTED WOODWORK

Preparing the Surface

REMOVING DEFECTIVE PAINTWORK

Paintwork which is blistered or cracked should be removed. Although it is possible to cut out the blisters and carefully prime, stop and fill the scars and cracks before repainting, such procedure usually costs more than completely stripping the surface and very rarely gives satisfactory results. Paint can either be removed with solvents or by heat and the choice of method will largely depend upon the type of surface.

Solvents are of three types: alkaline, such as caustic soda or potash mixed with starch or flour; volatile solvents controlled by waxes; and solvents of low boiling point. Caustic or alkaline removers are best used on non-absorbent surfaces such as glass, tiles and metal, as they are difficult to wash completely from porous materials and tend to raise the grain of wood. After the removal of the paint, repeated washing with vinegar and water to neutralise the alkali is necessary and the surface allowed to dry thoroughly before any paint is applied. Spirit removers are much better than burning off on such surfaces as metal and plaster, which quickly conduct the heat away. Some removers are highly inflammable and due regard should be paid to fire risks. Adequate attention should be paid to ventilation during their use. A full generous coat should be applied and left on the surface until the paint softens and expands, when it can be scraped off and put in a container. Caustic soda is dangerous to flesh, clothing, fabrics, linoleum, and must be effectively controlled. Spirit solvents are not so dangerous but should be used carefully. When paint is removed with a spirit solvent, the surface should be washed with white spirit and rubbed with steel wool to prepare the surface for painting, which can then be done without delay.

Burning off can be done with a petrol or paraffin burning blowlamp or gas torch. The burning off should be thorough, as any charred paint left on the surface will have lost its elasticity and prove detrimental to the new paint; but scorching of the wood should be avoided, as charred wood will not hold paint well. Burning-off should be done on mouldings and enrichments first and then all flat surfaces. As heat rises, the burning off should be started at the bottom of each surface and the stripping knife or shave-hook moved just a little behind the flame. Shave-hooks will be required for mouldings and they should be kept clean and sharp. If force is needed to remove the paint, the flame has not been on the paint long enough. When completed, the surface should be rubbed down and the surface washed with white spirit to remove any soot. As the paint is removed, it should be scraped from the knife into a container and afterwards removed from the scene of painting operations. If it falls on the floor it may set fire to or damage it; and, if left lying around, is injurious, since it may contain lead.

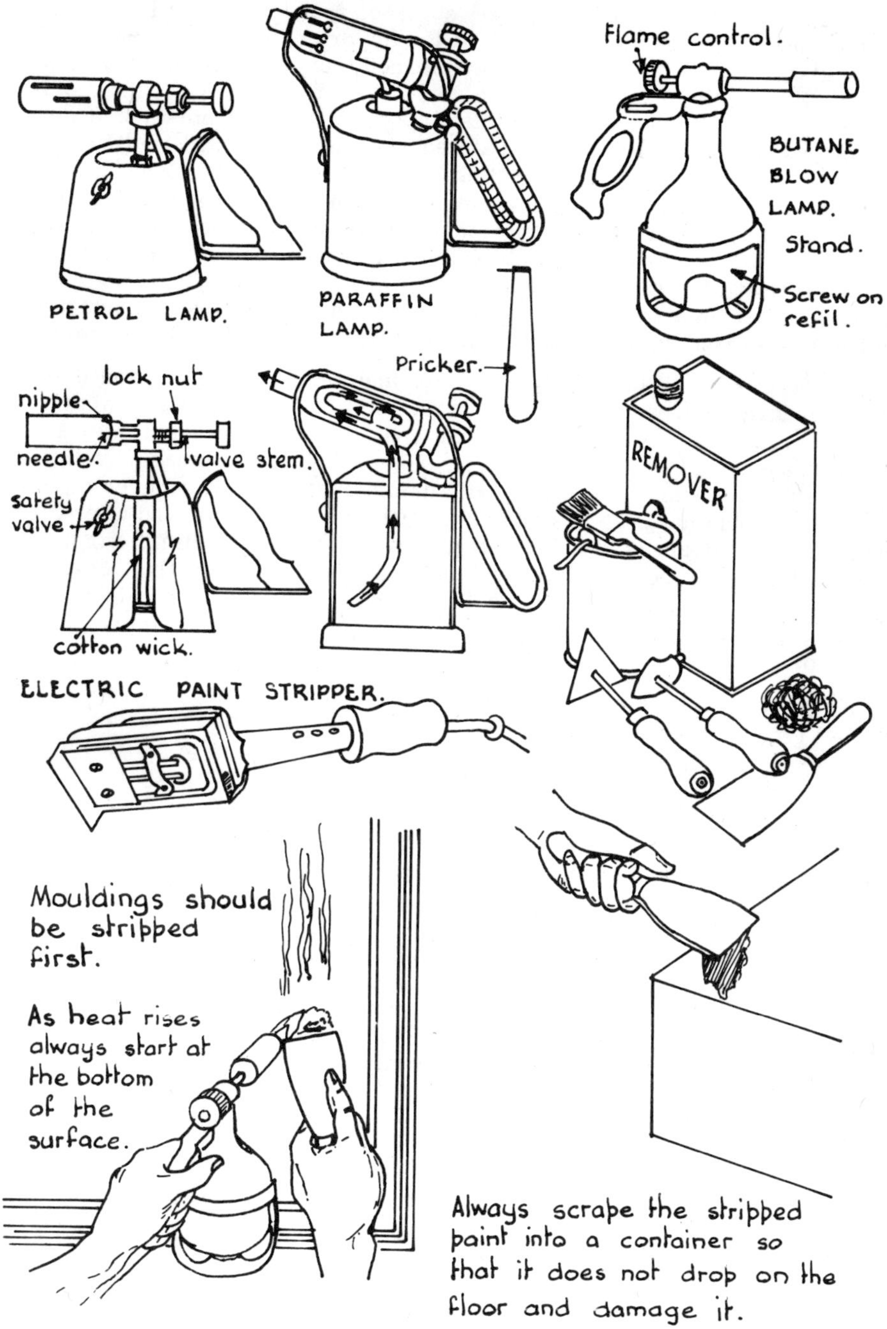

REMOVING DEFECTIVE PAINTWORK.

Preparing the Surface

IRON AND STEELWORK I

The problem of protecting iron and steel surfaces from corrosion is a major one; but, of the various methods suggested for the protection of iron and steel by paint, all agree that the metal must be cleaned thoroughly before any paint is applied. Rust, which results from the combined attack of oxygen and moisture, is a porous substance which acts as a sponge and holds moisture at the face of the metal so that even if it is painted over the attack still continues and the paint film is eventually pushed off. More dangerous than rust is mill scale, consisting of compacted areas of metal which form on the surface as the metal is milled. These areas are more dense than the main body and, due to the difference of structure, minute electric cells are set up over the surface, with the result that the mill scale becomes cathodic and is unaffected, while the main body becomes anodic and is eaten away. It is therefore important that all mill scale and rust should be removed before any paint is applied.

There are several methods of preparing iron and steel for painting and the one adopted will largely depend on the size of the job. On domestic work, where the areas of metal are usually small, being confined to gutters, rainwater and stack pipes, sashes, gates and railings, vigorous rubbing with emergy cloth, steel wool or wire brush, or chipping with hammer and chisel, scraper or file, often suffices. Obstinate patches can be heated with a blowlamp and afterwards scraped off. Larger areas in industrial plants will need more drastic treatment according to the condition of the surface and amount of rusting. Scraping and chipping with tungsten carbide steel scrapers or hammer and chisel are suitable methods for isolated areas, but whole structures may need pneumatic hammers or portable electric sanders or grinders.

Flame cleaning with special burners using oxy-acetylene leaves rust as a loose powder on the surface, loosens mill scale, burns off the paint and leaves the surface warm and moisture-free long enough to enable a coat of primer to be applied over it. When using the flame, it is necessary to wear goggles, leather gauntlet gloves and, when working in a confined area, a face mask against fumes from old paintwork.

Grit blasting, which involved the use of special appliances, is a better method of treating structural work before erection, since the size of the appliances limits their use on structures in situ. The abrasive material of cast iron grit, aluminium oxide or carborundum is projected at the surface by compressed air, and a high pressure ejector system picks up all abrasive and dust from the cleaned surface. It is very important to remove all particles of rust and dirt from the prepared surface; otherwise galvanic action will take place again, and in this respect grit blasting coupled with vacuum removal of debris is a distinct advantage.

To convert kN/m^2 into lbs per sq. in. divide by 7.
This gives an approximate conversion.

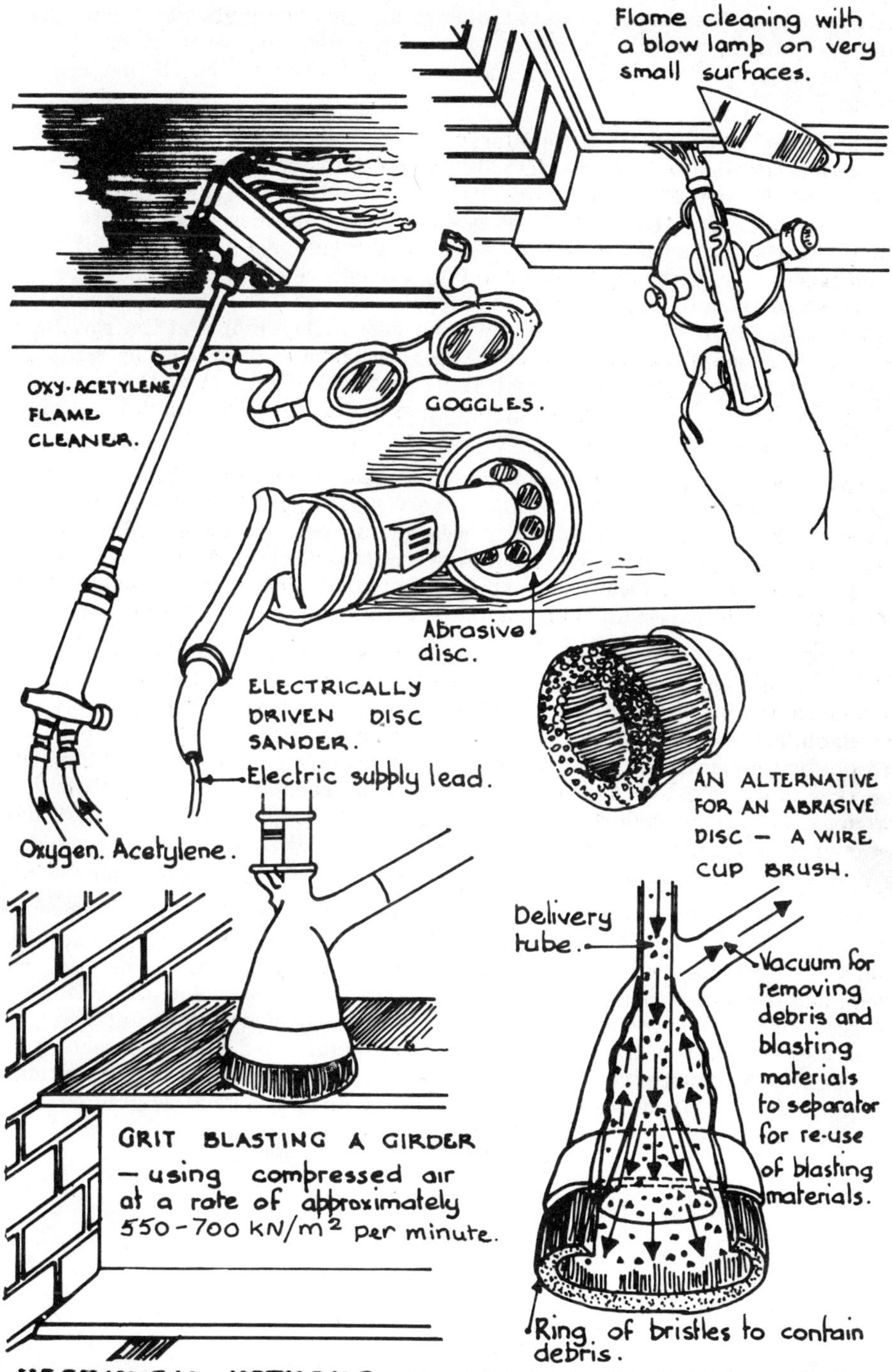

MECHANICAL METHODS OF PREPARING IRON & STEEL.

IRON AND STEELWORK II

Other methods of treating iron and steelwork are by chemicals which have the advantage that they do not require expensive equipment to operate them. In most cases a paint kettle and brush and the usual abrasives are all that are necessary.

There are several proprietary liquids, but all aim at coating the face of the metal with a substance which either will not rust so quickly as the untreated metal or stifles rust should it occur. Those liquids based on phosphoric acid give an iron-phosphate film and those on tannic acid produce iron tannate. Tanning liquors are not harmful to the skin or clothing and do not give off noxious fumes. They should be applied with a brush in a thin but continuous film, since spraying or dipping gives very poor results. The tanning rapidly takes up oxygen and presents the metal receiving it, developing at the same time a weather-resisting film of blue-black iron tannate. For treating mild corrosion, phosphoric acid can be used. It is comparatively mild and does not require neutralizing. After the coating has been on for about 20 minutes the surface can be wiped dry and painted straight away.

Where corrosion is advanced a stronger treatment is necessary and stronger acids such as sulphuric are used. This involves rinsing the steelwork after treatment and the wearing of rubber gloves and aprons, and eyeshields. Suggested treatments for new work: (a) Remove mill scale with descaling jelly; (b) Wash down with clean water; (c) Apply passivating liquid or phosphoric acid and leave for 20 minutes; (d) Wipe surface dry and apply chemical sealer.

On previously painted ironwork, it may be necessary to remove the paint with a paint solvent; caustic soda is cheap and effective so long as the surface is well washed and neutralized afterwards. Spirit solvents are more expensive but reliable and safer in action. If the whole of the surface can be stripped, so much the better, since the paint system can be brought forward as a whole, but if the cost is too great then only the areas of and around the rusting will have to be done, and the procedure will be as follows: (a) Strip old paintwork; (b) Rub down corroded areas with wire brush or steel wool; (c) Apply passivating liquid and leave for 20 minutes; (d) Wipe surface dry and apply chemical sealer.

Where only selected areas have been stripped, these should be primed and brought forward in a satisfactory manner before the general painting is commenced. To help cut the cost, some firms supply materials which seal the surface and form a chromate coating on it at the same time. Brushes used in these materials should be cleaned immediately after use, and the manufacturer's instructions carefully followed. When cleaning metal work, special attention should be given to bolts, rivets and nuts, since it is around these areas that rusting is likely to occur. When the surface has been cleaned, it should be primed the same day and not left to weather overnight.

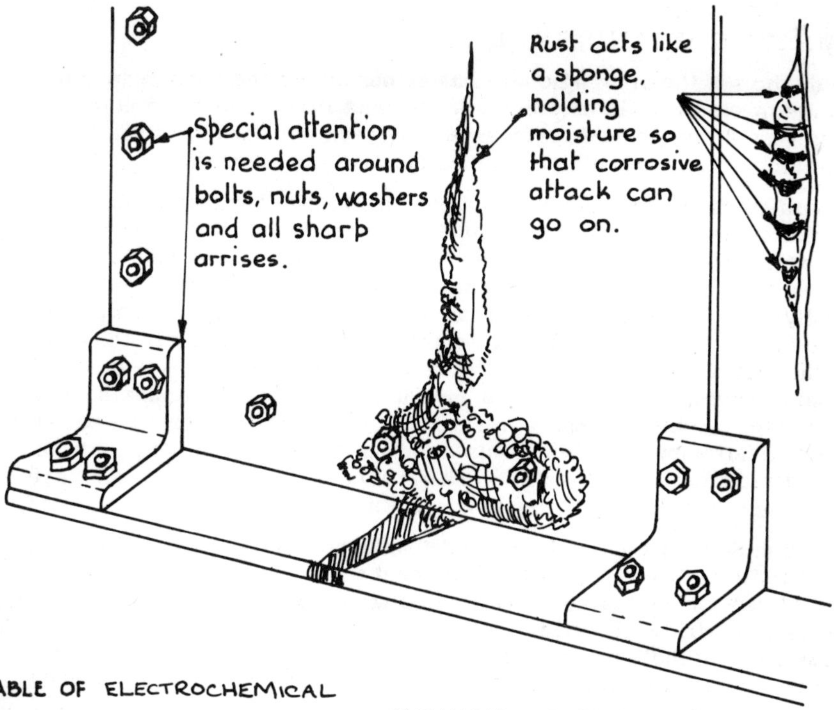

TABLE OF ELECTROCHEMICAL ACTION.

Anode +
Magnesium
Aluminium
Duralumin
Zinc
Cadmium
Iron
Soft solder
Tin
Lead
Brass
Copper
Antimony
Cathode −

CHEMICAL PREPARATION OF CORRODED IRONWORK.

1. Descaling Jelly.
2. Washing with clean water.
3. Passivating liquid.
4. Sealer.
5 Primer.

These can effectively be used on construction work where it would be impossible to use mechanical equipment.

ELECTROCHEMICAL ACTION & CHEMICAL PREPARATION OF SURFACES.

PRIMING IRON AND STEELWORK

The selection of the priming coat is as important as the cleaning of the metal, which in any case should be primed as soon as it is cleaned. In no circumstances should the cleaned surface be left overnight, but if this is unavoidable the cleaning process should be carried out again the next day.

From time to time claims have been put forward for various types of primers, and, while some manufacturers produce primers which merely act as sealers, others rely on the reactions between the pigment and the metal to bring about a passive state of the metal, so stifling or inhibiting rust. The sealer types are generally graphite or red oxide and rely solely upon the film formed on the surface for shutting out air and moisture; but the best types are those which stifle the corrosive tendencies of the metal. These primers are based on zinc chromate, chrome greens, red lead, metallic lead or calcium plumbate, with a medium composed of two parts linseed stand oil and three parts alkali-refined linseed oil. Calcium plumbate paints do not blister to the same extent as other paints and are useful for marine work. Paints based on epoxy resins are also favoured as they have good resistance to both alkali and corrosive action.

The British Standard primer for steelwork consists of a mixture of equal parts of white and red lead with 25 per cent their combined volume of asbestine. One of the advantages of red lead applied with a brush is that it will mix with any moisture left on the surface of the metal and allow it to evaporate in the same manner as the thinner while the paint adheres firmly to the surface, which makes it very suitable for foundry coats. If, however, some time must elapse between the application of the priming coat and the rest of the painting, zinc chromate is better than red lead, since it is more durable.

Priming paints used on metal should spread easily under the brush; dry throughout and not merely skin on the surface; and form an elastic film which will adhere to it. Cheap primers have inferior wearing qualities and afford poor protection. On previously painted work, where bare patches occur it is essential to give such areas the same careful treatment as new metal. The primer should be just as carefully selected and the paint system brough forward to match the general condition of the surface before any of the main painting is done. In the same way, all rivet heads, bolts and sharp edges should receive an extra coat of paint, since it is on these parts, due to their shape, that the paint is thinnest. For cheap work, where the finishing colour is of little consequence, use can be made of bituminous paints, which are particularly suitable for underground ironwork or against corrosive action by tidal waters. They are also generally resistant to acids and alkalis.

Self-etch primers are half way between paint primers and phosphate pre-treatment; they consist of polyvinyl butyral resin pigmented with zinc chromate, with phosphoric acid solution added just before use.

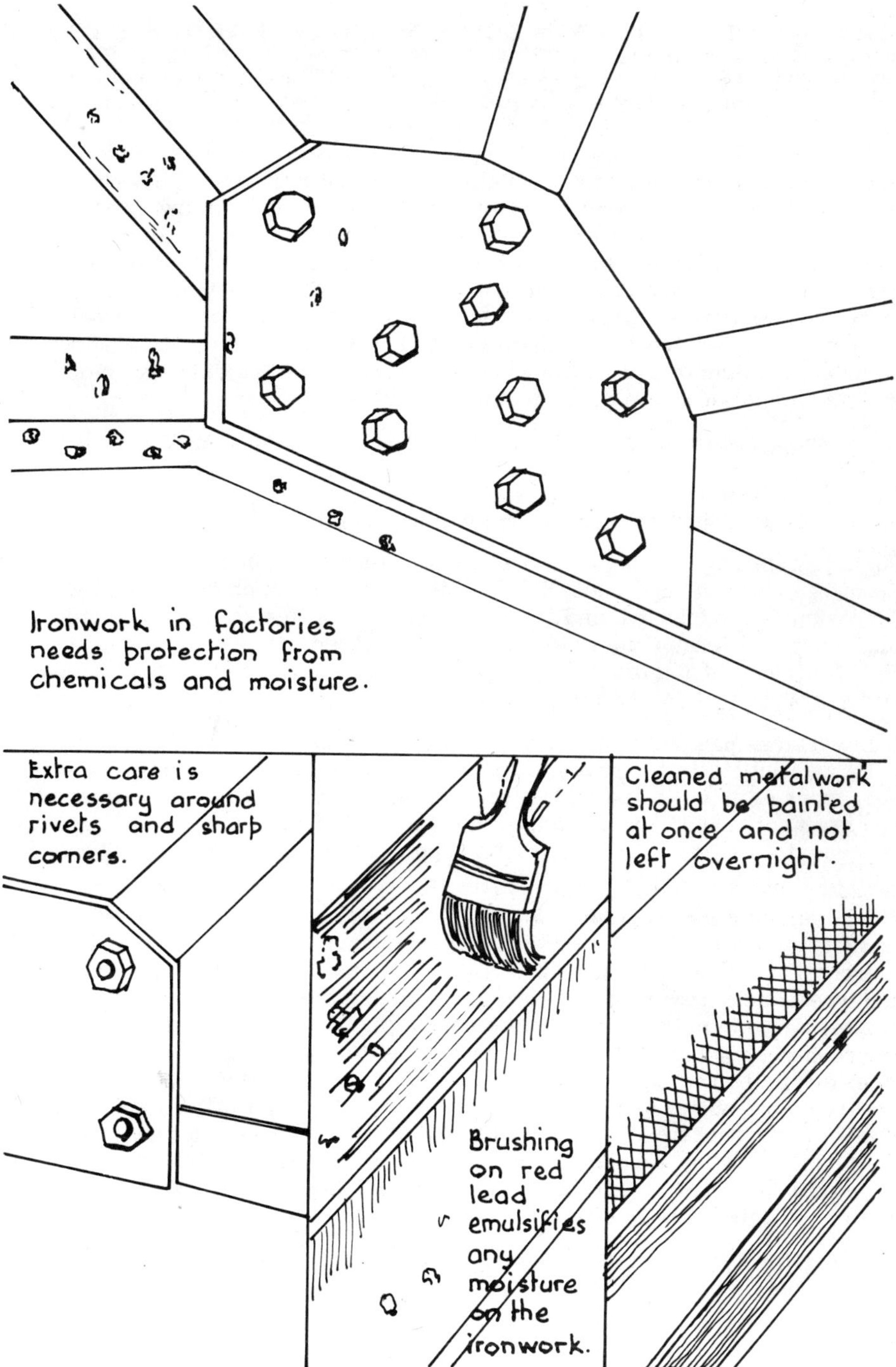

PAINTING IRON AND STEEL.

TREATMENTS AND PRIMERS FOR NON-FERROUS METALS

Non-ferrous metals do not corrode so rapidly as iron and steel and in some cases the film of corrosion itself prevents further attack taking place. The coating of green verdigris which forms on copper roofs is a good example. Zinc tends to acquire a white rust deposit of combined zinc oxide and zinc carbonate on its surface. Units which have been pre-treated by sheradizing or zinc spraying can be painted straight away and do not require any further preparation than making them dust and grease free. Untreated zinc and the zinc coating on galvanized iron should be cleaned and etched. Washing with sugar soap, pumice soap or detergent will remove all grease and slightly etch the surface, which can further be 'keyed' with emery cloth or steel wool. Pretreatments consist of phosphate or chromate passivation. Most paints are suitable for galvanized iron and zinc; red oxide or well-thinned exterior enamel tinted with red oxide are both recommended.

Cadmium coated steel should be treated in just the same way as zinc, but weathering is not considered suitable. For priming terne plate, which is iron sheeting covered with a lead and tin alloy, no special preparation is necessary and all types of primers, except graphite, are suitable.

When magnesium and magnesium alloys are exposed to moisture or damp air a coating of non-adherent alkaline hydrated magnesium oxide is rapidly formed. Corrosion is progressive and the metal must be protected at every stage. Magnesium and magnesium alloys are best cleaned with a 5-10 per cent solution of nitric acid and primed with zinc chromate. Self-etch primers, lead paints, graphite, aluminium or bronze paints are not suitable for these surfaces.

Lead surfaces can be treated with a phosphate solution, or abraded, using white spirit as a lubricant to prevent particles of poisonous metal flying in the air. Most types of paint are suitable, except graphite. Copper is best primed with a varnish media based on oil modified alkyd resin. Some authorities suggest that where a gloss finish is required on a copper or lead surface two coats of gloss paint are all that is required.

For aluminium and aluminium alloys, cleaning and etching are essential as the surface quickly becomes covered by a tenacious self-healing oxide film which prevents further attack but makes the reception of paint poor. One recommended type of pre-treatment is 'Alocrom' which can be power-sprayed to work in situ. Suitable primers are self-etch and zinc chromate. Red and white lead or graphite should not be used. Where two different metals abut, both touching faces should be coated with two coats of bituminous or rubber paint before they are fixed, to insulate one metal from the other. Where a metal face touches acid-bearing woods such as chestnut, oak, or alkaline-bearing materials such as cement or concrete, both faces should be coated with two coats of bituminous paints before fixing. No priming should be done on metalwork in damp weather or late in the day so that condensation can take place on the surface.

All other metals should be cleaned and the surface 'keyed' by rubbing with steel wool or emery cloth.

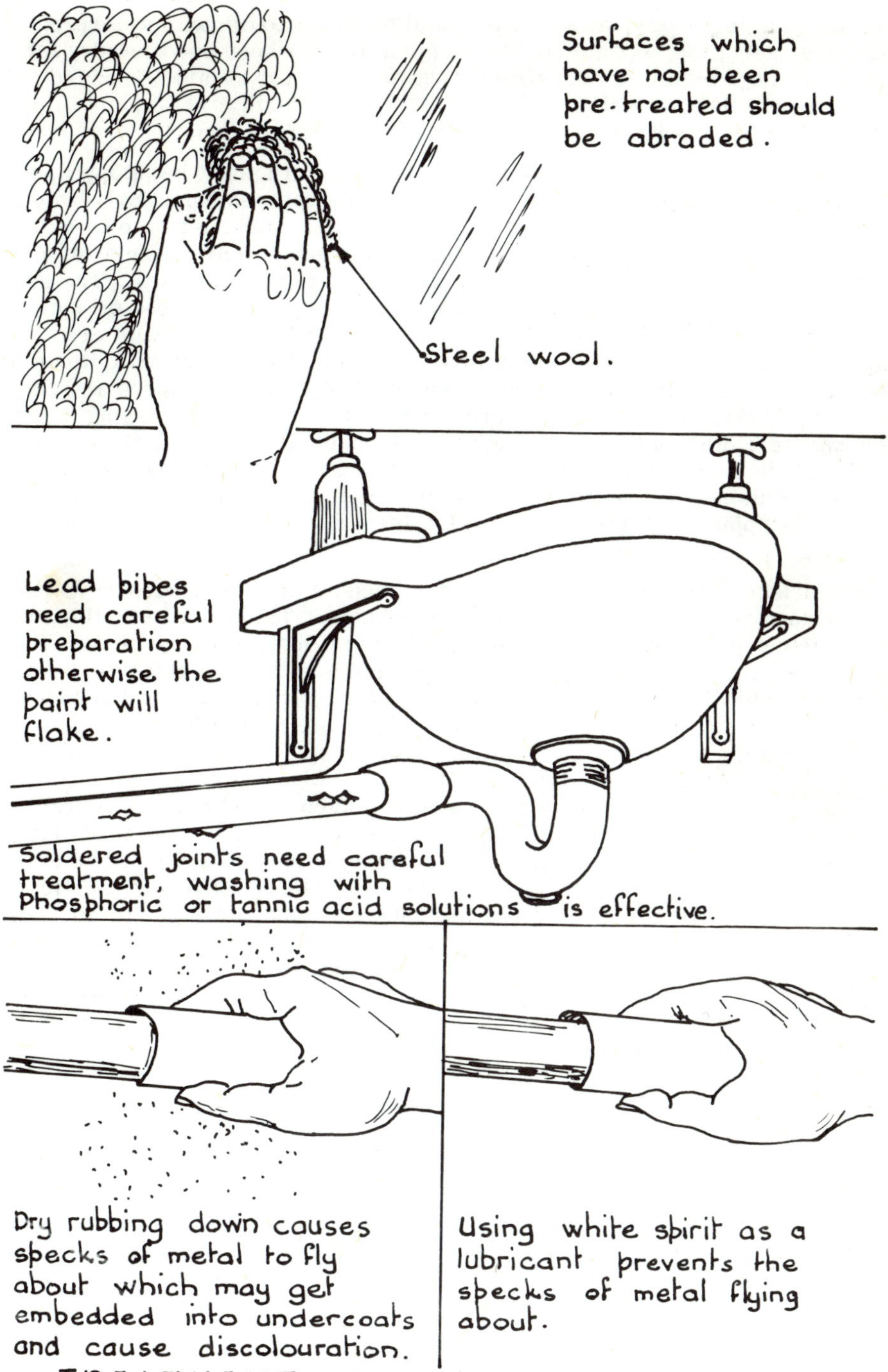

TREATMENT OF NON-FERROUS METALS.

Stopping and Filling

When the priming is dry, the surface should be rubbed down with waterproof glasspaper using the 'wet method' if lead priming has been used, but dry for other types of priming. It is desirable that the next coat should be applied without undue delay, since this assists the satisfactory build up of the paint film, but not before the necessary stopping and filling has been done.

All open joints and cracks, especially with outside work, must be 'stopped up' or 'made good' to prevent rain penetration. Joints between brickwork and door or window frames are best done with mastic, using a caulking gun to force it well into the opening; otherwise, pointing with a cement, lime sand mortar in the proportions of 1/1/6 is advised, the joint being well wetted first. The usual stopping for woodwork is linseed oil putty or equal parts of linseed oil putty and paste white lead. For the best work, paste white lead mixed with dry sifted whiting and gold size is used, with a little red lead added. If too stiff, it can be reduced with one or two drops of white spirit. Proprietary stoppers are also available. The stopping should be forced well into the hole by pressing the knife over half its width and then pressing the other half. The stopping should completely fill the hole and any surplus material on the face should be carefully removed. Fine plaster mixed with undercoating to a stiff plastic condition is useful for stopping, and when mixed with a little more undercoating is useful for filling.

Where surfaces are uneven due to age or misuse, it is often necessary to fill up the indentations to give a semblance of smoothness. The amount of filling done will largely depend upon the specification. Several materials can be used for filling; glue size and whiting or paste soft distemper and gold size are useful for cheap inside work. Thick washable distemper with a little varnish added or proprietary fillers are more durable but all distemper fillers should be coated with a glaze of varnish and white spirit to stop their suction. For outside work, paste white lead, whiting, gold size and white spirit or slate filler (according to the finishing colour) is more satisfactory. The filling is transferred to the surface on a flexible filling knife and applied in smooth strokes in the direction of the longest length of the surface or in such a manner as to deposit the filling in the position it is most needed. The art of filling is to leave the surface perfectly smooth with as few ridges as possible, so as to minimise rubbing down. On mouldings, the filling is thinned and applied evenly with a brush. If the filling gathers or rolls up during application, it is too thick and needs thinning with water or white spirit, according to the material. Distemper based fillers can be lightly rubbed down dry with fine glue bound glasspaper; slate and hard fillers need wet rubbing down.

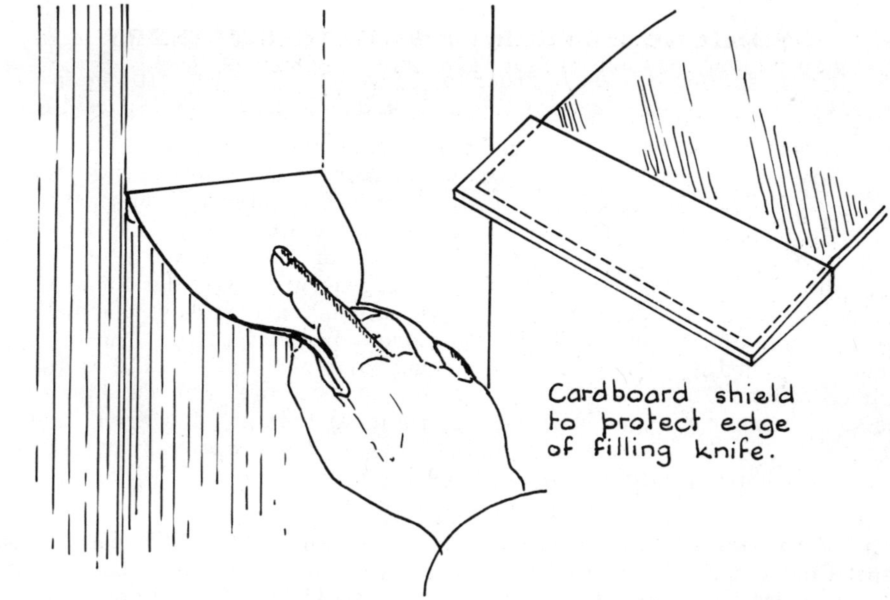

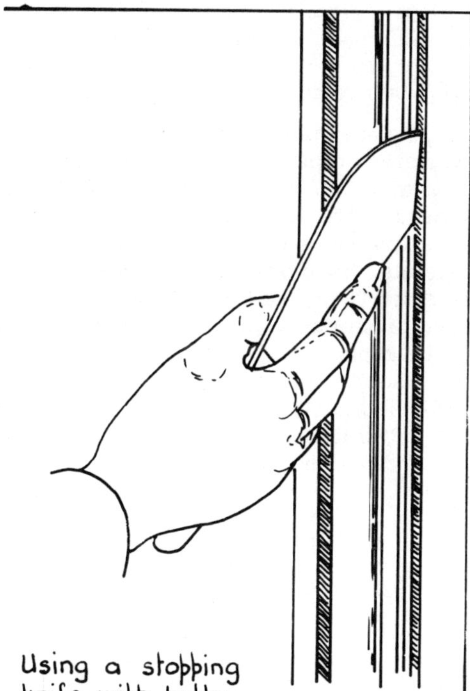

Using a stopping
knife with putty
or hard stopping.

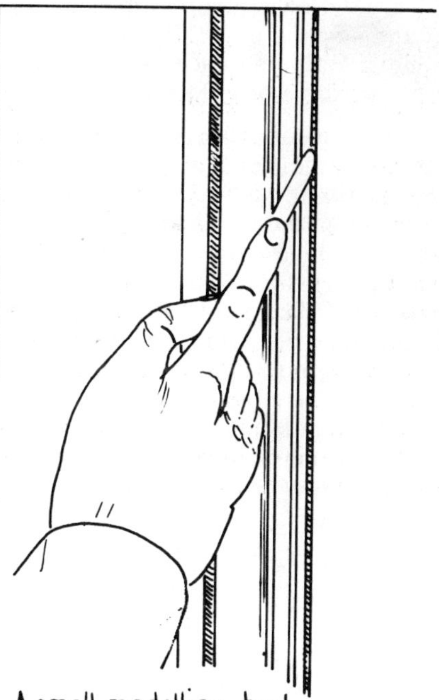

A small modelling tool
or quirk stick is useful
as an alternative to a
stopping knife for small cracks.

STOPPING AND FILLING.

Staining

AN ECONOMICAL METHOD OF TREATING JOINERY

Stain is applied to woodwork in order to change its colour but still retain the natural grain of the wood, and for this reason they should be transparent but strong.

There are several types of stain used by the painter, the chief being: (a) Water stains such as potassium permanganate dissolved in water; or transparent pigments such as siennas, umbers, Prussian blue, crimson, gamboge in water thinned with glue size. They are very cheap and of good colour but tend to raise the grain of the wood. (b) Spirit stains, generally in powder form, which are mixed with methylated spirits. These materials are quick drying and need speedy application if satisfactory results are to be obtained. (c) Transparent pigments ground in oil and mixed with equal parts of linseed oil and white spirit. (d) Ready made stains, e.g. 'Matsine' or 'Grainite', which are thinned in the same manner as pigments in oil. Types (c) and (d) are slow drying and easy to handle. They have slight protective value. (e) Proprietary preservative stains, which are usually based on phenols or naphthalene and are formulated as a protection against woodworm and fungi (dry rot).

The wood work should be prepared by rubbing down with fine glue-bound glass-paper (Nos. 0 and 1,) rubbing always with the grain so that the surface is not disfigured by scratches. If the woodwork is discoloured, it should be bleached with a solution of oxalic acid or a proprietary wood bleach left on overnight and the surface carefully rubbed down as before. End grains like tenons take up more stain and become darker in colour and it is usual to coat these areas with a mixture of gold size and turpentine before staining.

The stain should be applied with a brush, covering all the flat surfaces first and the mouldings last. For the best work the stain should be left on the surface for 5 min and then any surplus stain wiped off with a piece of soft rag, so that an even colour is obtained. For general work, it is usual to thin the stain to the required colour by testing it on a spare piece of wood of similar texture, and then brushing it on as evenly as possible. Hard woods with a coarse open grain, like some types of oak, should be filled first; and this can best be done by knifing over the surface a thin mixture of plaster and water, allowing it to dry and rubbing down the surface with No. 0 glasspaper before staining. Proprietary fillers are also useful for this purpose. When the stain is dry, any necessary stopping should be done either with a proprietary stopper or putty coloured with oil stainers to match the general effect of the stained surface. Stained surfaces need protecting from constant handling and the usual methods of finishing are spirit polishing, gloss varnish, flat varnish and wax polishing.

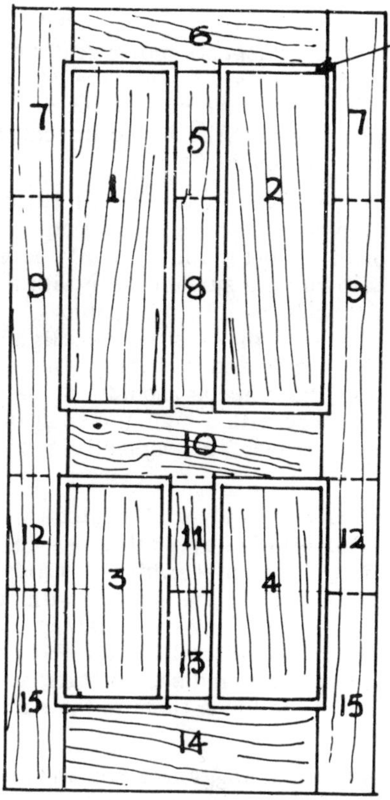

Panel mouldings should be stained last.

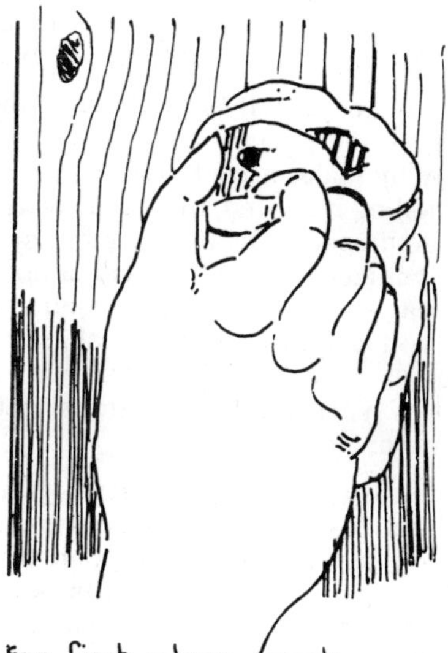

For first class work, after the stain has been applied remove the surplus with a piece of lint free rag.

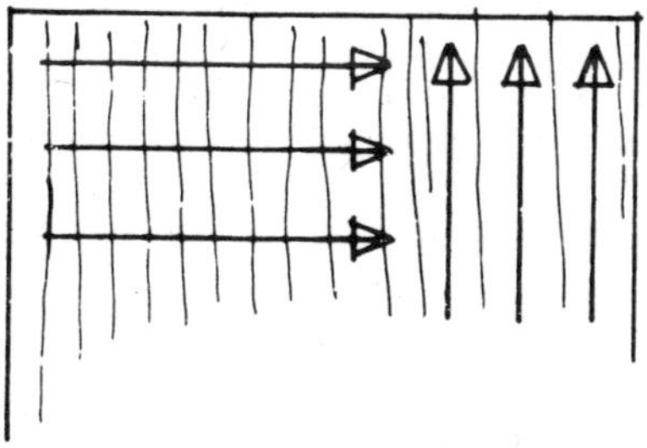

Do not glasspaper across the direction of the grain, otherwise unsightly scratches will spoil the surface.

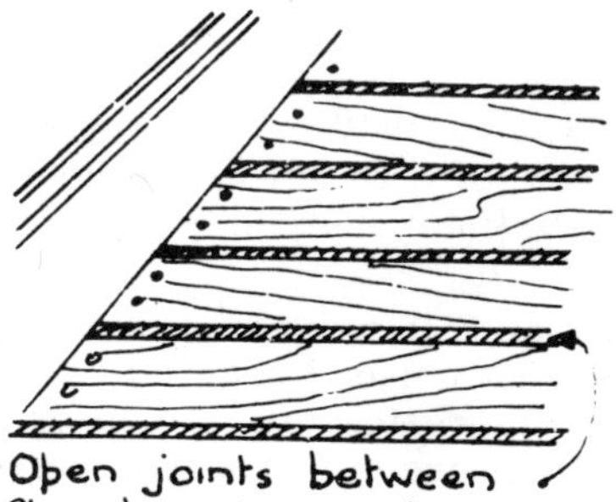

Open joints between floor boards can be filled with thin strips of wood or proprietary wood fillers.

STAINING WOODWORK.

Varnishes and Varnishing

Varnishes are transparent coatings which are used for the protection of new and stained woodwork, also new paintwork, signs, wallpaper, floors and furniture. They are made by cooking resins with oils and driers. The oils are chiefly linseed and tung oil and the resins both natural and manufactured. Natural resins, termed copals, are either dug from the ground (fossil) or are tapped from living trees (recent). Synthetic resins are manufactured; they produce varnishes which are paler and in some respects more weather resistant than natural resin varnishes. Frequently varnishes are made which contain both natural and synthetic resins, resulting in varnishes which are of finer quality than those made with only one type of resin. About 60 different types of varnish, including gold size and black japan, are made.

The surface which is to be varnished must be clean, hard, and with little or no gloss. If the surface is glossy or oily in any way the varnish will not brush on easily or make a continous film. For work finished in gloss or oily paint, grained or marbled, the usual practice is to rub the surface over with a damp sponge dipped in whiting, or moistened with white spirit. Wallpaper which is to be varnished should be given 2 coats of glue size, the first coat applied hot with vertical strokes, and the second applied when cold and laid off horizontally, both with a minimum of brushing. Washable distempers should also be treated in the same way. The majority of varnishes dry with a high gloss but there are two types which dry almost without sheen; these are eggshell flat and dead flat varnishes.

For the best work, two coats of varnish should be applied. For gloss varnishes, the first coat can be thinned with turpentine, followed by a full gloss coat. Another method is to apply one coat of 'flatting' varnish, which is a gloss undercoating varnish, and when this is dry and hard, to rub the surface with pumice powder and water on a felt pad to dull it before applying the finishing coat. For flat varnishes, an undercoat of gloss varnish is beneficial and the flat varnish is best stippled. Varnishes are sensitive materials and should stand at least half an hour in the room they are to be used in to bring them to the same temperature as the room.

Varnishing should not be done in very cold, wet or foggy weather and not too late in the day, otherwise it tends to bloom—turn milky or whitish. It is likewise unnecessary to damp floors before varnishing. When varnishing, very great care should be taken to ensure that all brushes and kettles, the surface and surroundings, are perfectly clean. No pains should be spared in this direction. The varnish should be applied firmly with a minimum of brushing. Varnishes improve with age and the varnish which has been in stock the longest should be used first. On no account should varnishes be shaken up, otherwise the finish will be 'bitty'.

Emulsion Finishes are transparent materials which produce a flat or matt surface and are useful over pale and delicate surfaces such as wallpapers. Based on acrylic resins, they are odourless, easy to apply by brush, roller or spray, and dry in one to two hours. Pre-treatment is unnecessary.

Varnishing should not be done too late in the day.
GO
HALL STORES
Always cover the kettle & brush with paper during breaks.
Varnishes should be stood in the room half an hour before use.
VARN
Flat varnishes should be stood in a bucket of hot water.
Fine water proof glasspaper
or pumice powder on moistened felt pads are suitable for rubbing down.
Varnish from an opened can should be strained through fine muslin or a piece of silk or nylon stocking.
FLOUR
A varnish surface can be polished with a soft duster and flour.

VARNISHING.

Special Effects

SCUMBLING, GLAZING AND SHADING

These three processes are slightly different from plain painting, as they colour the surface but do not hide it, and any one process is an addition to and not a substitute for a coat of paint. They consist of applying translucent materials over a painted surface and afterwards manipulating it with brushes, combs, pieces of cloth, etc., to produce textured effects. The ground needs to be hard, preferably with a semi-gloss finish so that the material can be brushed on and worked over the surface evenly and easily. If the ground is soft, flat, or not properly dry, it will be difficult to obtain an even effect.

Ready made scumbles can be obtained which only need thinning with white spirit. If the material drags or dries too quickly, 2 or 3 drops of linseed oil should be added. When thinned, scumbles are referred to as glazes. Effective glazes can be made up with a medium of equal parts of linseed oil, varnish and white spirit, and small quantities of undercoat and transparent pigments, such as yellow ochre, raw and burnt sienna, raw and burnt umber, zinc yellow, ultramarine and monastral blue, crimson, and vegetable black. The brushes and kettle need to be perfectly clean, as this type of finish shows every defect.

The glaze should be applied evenly so as to minimize work with the stippler or softener. The usual finishes are fine, even textures produced with hair stipplers, coarser textures produced with different grades of rubber stipplers, and varied textures made with rag, wash-leather, paper, etc. After application, the glaze should be left on the surface for a few minutes to settle before any use is made of the stippler, and the stippling should be done as systematically as possible. The stippler should not be used in straight lines but the direction of the stippler changed frequently, travelling in narrow bands from side to side. For stippling in internal angles, use can be made of a single row stippler or a good quality dusting brush, and it is much better to stipple the corners and angles first before the whole flank.

After use with glazes, hair stipplers should be gently rinsed or wiped with white spirit and either dabbed on scrap paper or carefully wiped before being washed with soap and water and hung up to dry. Rubber stipplers should never be rinsed in white spirit since it soon perishes the rubber; they should be carefully washed with soap and water and dusted with French chalk or talc before being hung up.

Where a graduated effect is required, glazes of three or more strengths are required. Each glaze should be applied as evenly as possible and where the different glazes join they should be blended with the brush before being stippled. Textures obtained by rolling a piece of wash-leather, soft cloth or paper over the wet glaze are termed rag-rolled. As the cloth touches the wet glaze it pushes it aside and produces patches of the glaze and ground colours side by side. The rolling should not be done in straight lines but in all directions. When finished with, the rag should not be screwed up tight but laid out flat until it can be disposed of, as the oil in the glaze can generate sufficient heat to cause spontaneous combustion. By applying 2 or 3 different coloured glazes over a broken white ground, many interesting colour effects can be obtained.

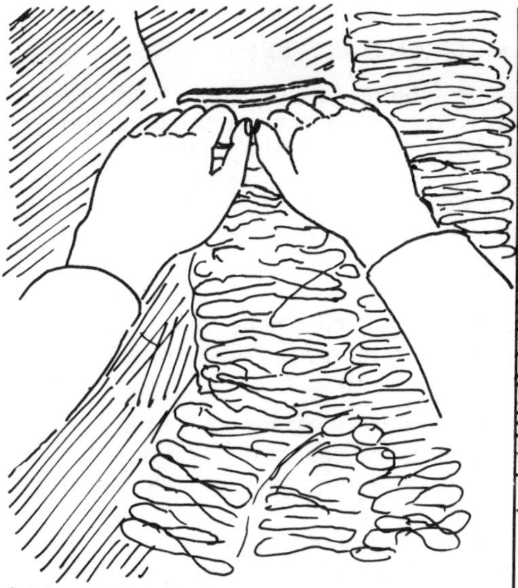

RAG ROLLING.
Start at the bottom of the surface and roll in all directions, not in straight lines.

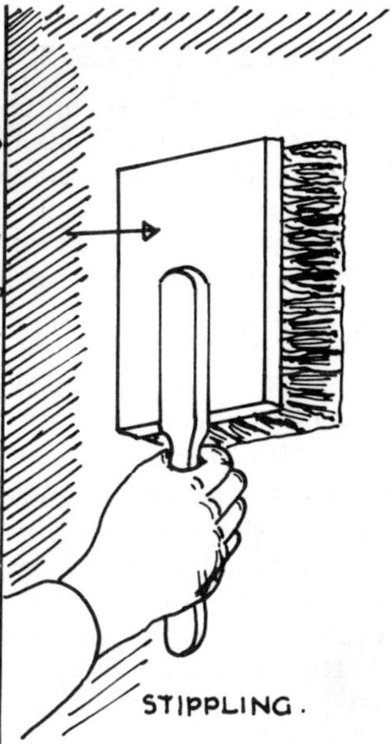

STIPPLING.

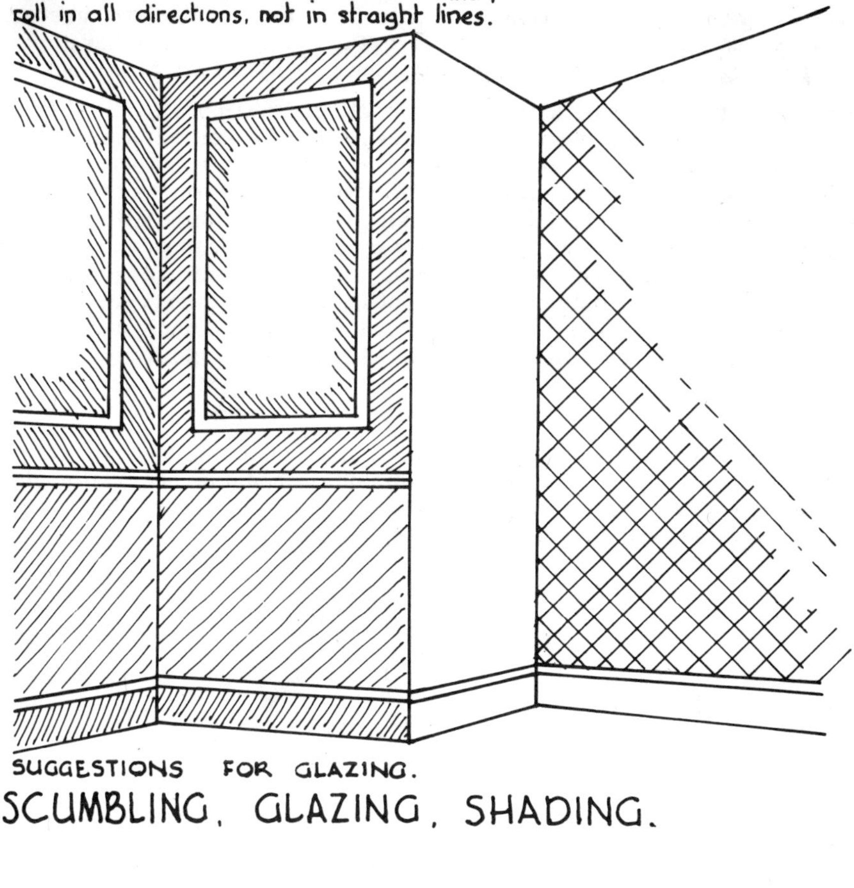

SUGGESTIONS FOR GLAZING.

SCUMBLING. GLAZING. SHADING.

Special Effects

SPONGE STIPPLE AND TEXTURE PAINTS

Other texture effects can be obtained by dabbing the wet glaze with a sponge dipped in white spirit or with various colours on a dry painted or distempered ground colour. In the former case, the spirit 'opens out' the glaze and reveals the ground colour, and the effect can be softened by using a softner in all directions. The sponge chosen should be 'open' and not too fine. For sponge stippling, the colours used should be soft rather than garish, and more or less of the same tone. The general method is to put a little of the colour on a palette or piece of board and transfer the colour to the surface with the sponge. An even all-over texture should be aimed at with each colour, so that the finished effect is regular and not spotty. Metallic paints are useful with this type of finish to give 'sparkle'.

Texture effects in low relief can be obtained by using plastic texture paints which can be used on previously decorated surfaces, bare plaster, building boards, or as a substitute for a setting or finishing coat on brickwork or cement. Plastic paints are in paste or powder form. The paste, being stiff and self-coloured, is a one coat process, but the powder is generally distempered or painted after it is dry. The powder has to be carefully added to water, stirring briskly all the time until a fairly stiff paste is obtained. For very fine textures or spray application, the material should be strained through a piece of coarse canvas. For small flanks, the material can be applied and then textured but where large areas have to be covered it is necessary to have one painter applying the paint and another following closely behind, texturing.

Methods of texturing include hair and rubber stippling, stippling with a flat piece of wood, rolling with a rubber or wooden roller, use of celluloid combs, pattern making with various painters knives, or applying the material and afterwards slapping it with the brush. Quite a number of textures are improved by passing a piece of celluloid dipped in water over them while still wet; otherwise it is better to lightly rub down the texture with M2 glasspaper when it is dry. The size content in the material not only binds it but kills any suction, so that it can be painted or distempered without any pre-treatment. A popular finish is to paint the surface in one colour and, when dry, to apply a glaze of a contrasting colour, wiping off the highlights with a piece of cloth or rubber squeegee. Plastic paints are best applied over surfaces which have a slight porosity: on shiny surfaces they tend to slide.

Where two or more different textures are required on a single surface, gummed brown paper strip can be used. The divisions should be set out, the gummed strip carefully applied and when the texturing has been done the gummed strip can be pulled away leaving clean sharp edges.

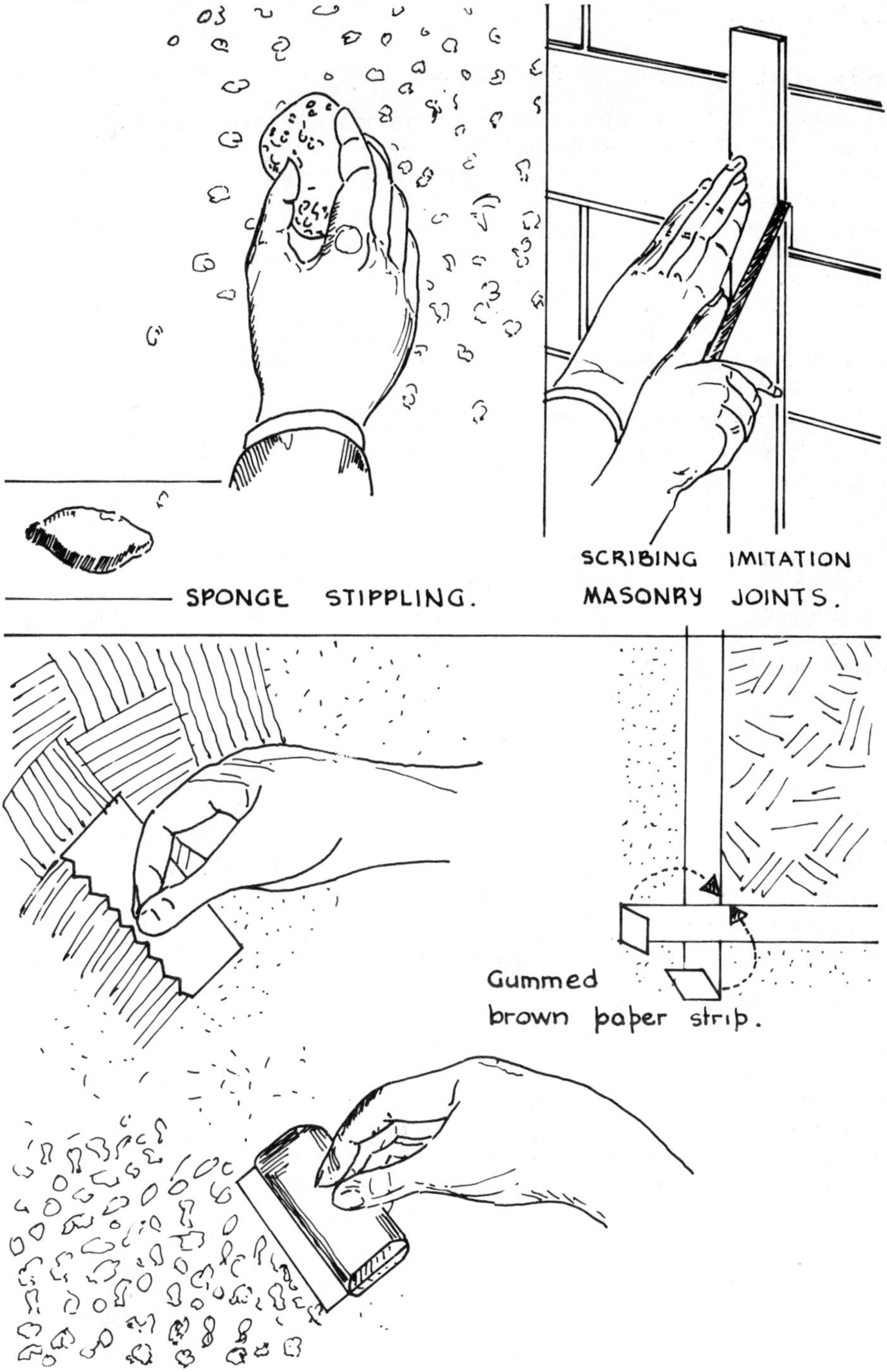

SPONGE STIPPLE & TEXTURE PAINTS.

Special Effects

BRUSH GRAINING

The grounds needed for graining are hard with a slight sheen. This allows the graining colour to be worked freely without biting into the surface. If the ground is too soft or oily 'cissing' will occur, which will prevent the graining colour being applied in a continuous film. To prevent 'cissing', the surface should be rubbed over with a damp sponge with a very little whiting or white spirit until all traces of sheen are removed. For graining, the colour of the ground represents the lightest colour in the wood, and it should never be made too bright, but a little on the greyed side.

Wood Effect	Colour of Ground
Natural Oak Ash and Sycamore Pine	White with ochre
Light Oak	Light buff from white, ochre, raw umber
Medium Oak	Dark buff from white, ochre, burnt umber
Dark and Pollard Oak	Dark buff with a touch of Venetian red
Weathered Oak	Aluminium or light grey from white, raw umber and black
Mahogany	Light Venetian red with white and orange chrome
Rosewood	Bright orange buff with orange chrome, white and Venetian red
Walnut	Dark tan with white, Venetian red, ochre and umber

The tools required are a 50 mm flat brush for brushing on the glaze, a brush grainer or dusting brush, and a board knife or joint rule for forming clean straight lines, and a piece of cloth. Any suitable woodlike coloured glazes can be used. Clear glazes are an advantage, since the same glaze suitably tinted can be used for a variety of woods. Useful graining colour can be made with equal parts of oil, white spirit and varnish, with a little whiting or undercoating added, and tinted with suitable stainers. The procedure should be the same as for plain painting. The glaze or graining colour should be applied evenly; any variation of depth of colour should be gradual, so that patchiness is avoided. When the glaze has been on the surface two or three minutes, the graining brush or duster is dragged along the surface, leaving streaks or lines of colour of varied width and depth, which can afterwards be lightly flogged or dabbed with the edge of the graining brush to produce a softer effect. The method of graining follows closely the construction of the joinery. Where the direction of two members abut: such as a muntin or rail and stile, a joint rule or broad knife is used to secure a knife-edge joint. This can be done by brushing and dragging one member, holding the rule in the exact position where the joint is required and then brushing in and dragging the other member. If desired, steel, leather or rubber combs can be used instead of the graining brush or duster.

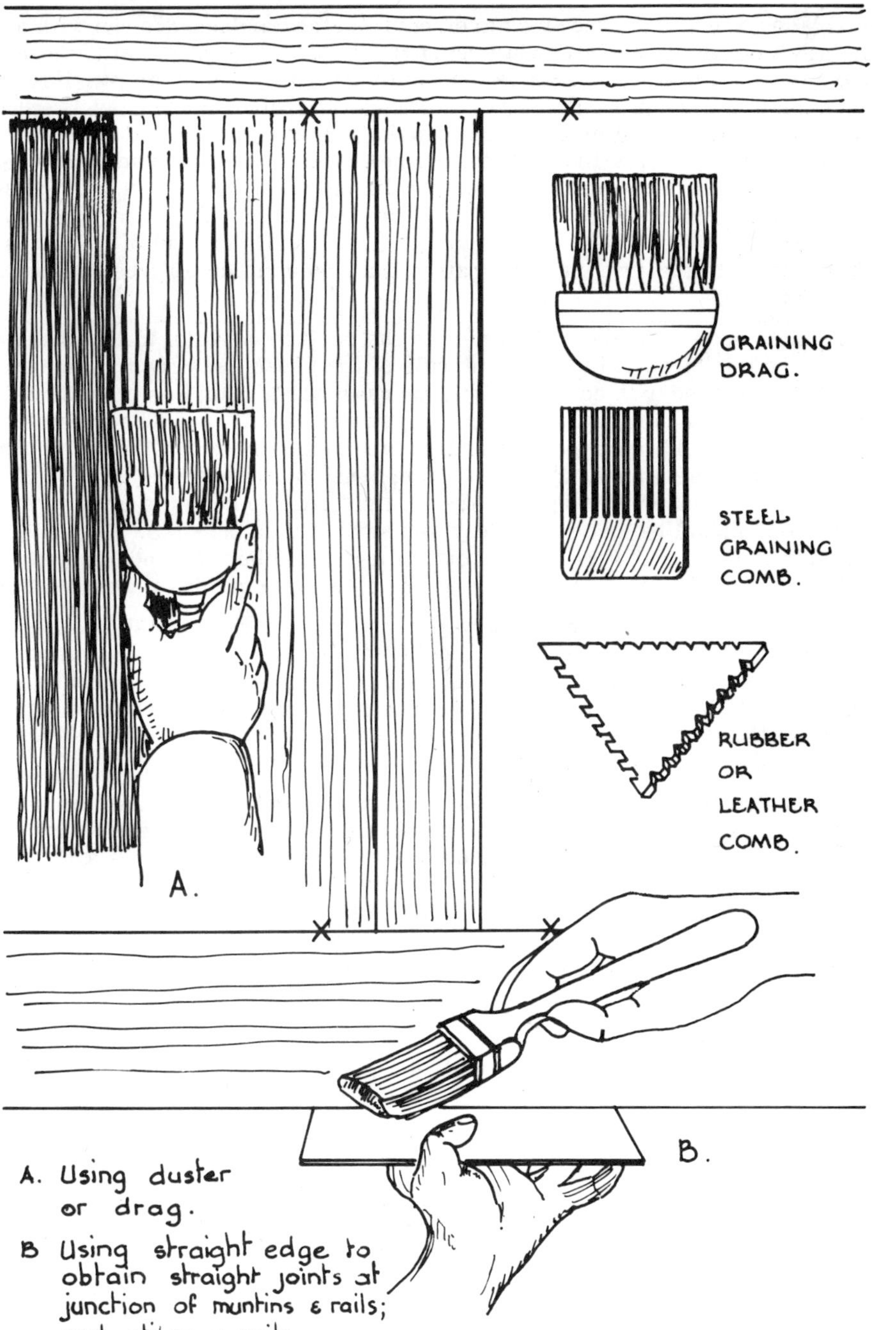

A. Using duster or drag.

B Using straight edge to obtain straight joints at junction of muntins & rails; and stiles & rails.

BRUSH GRAINING.

Running Lines

WITH FITCH AND WRITER, USE OF MASKING TAPES

In the cases where coloured lines are required, they are usually run with a lining fitch or writer and straight edge, each method requiring a slightly different technique. The lines should be correctly set out as lightly as possible, so that the setting out can easily be removed or obliterated. Setting out is usually done with a chalk line. Vertical lines should be made with the aid of a plumb-bob, and horizontal lines, like a dividing line at the top of a painted dado, checked with a spirit level. Where a quantity of lining has to be done, it is an advantage to mark a rod or piece of wood with the vertical heights, to save repeated measurements. Where ceiling lines and angles of walls are out of true, it is better to work to the existing lines rather than keep rigidly to true horizontal and vertical ones.

The straight edge is held bevelled side to the surface, parallel to the chalked line, but the thickness of the metal ferrule of the brush away from it, and the lining fitch charged with colour is lightly drawn along. For this operation the colour needs to be thin, but not too thin, otherwise it will run. The edge of the straight edge should be wiped frequently. When using a writer, the straight edge is held in the same manner and the writing brush between thumb and forefinger, with the fingers underneath against the straight edge to keep the brush steady. The strokes should be made as lightly and smoothly as possible. This method is similar to that used when picking out a single member of a moulding, the three fingers being used as a guide and run along the outside of the moulding.

Several straight edges are necessary, but the longest are the best to choose. For general work, 1 m is a suitable length, with smaller lengths for narrow spaces. Corners or ends should be put in with a writer. Where several rectangles have to be repeated, a quick method is to cut a small stencil and apply it as many times as corners are required. Lines can then be run from corner to corner.

Lines can also be made by sticking on each side of the width of the line self-adhesive masking or cellulose tape and painting in between. The masking tape can be used a second time; the cellulose tape once only, but it gives a much sharper edge. For producing panelled effects in plastic paint, after the setting out has been done, brown gummed tape the width of the line required is stuck along the lines, with an additional 20 mm or so in length folded back at the corners for ease of lifting. When the surface has been textured, the tape is removed, producing sharp divisions.

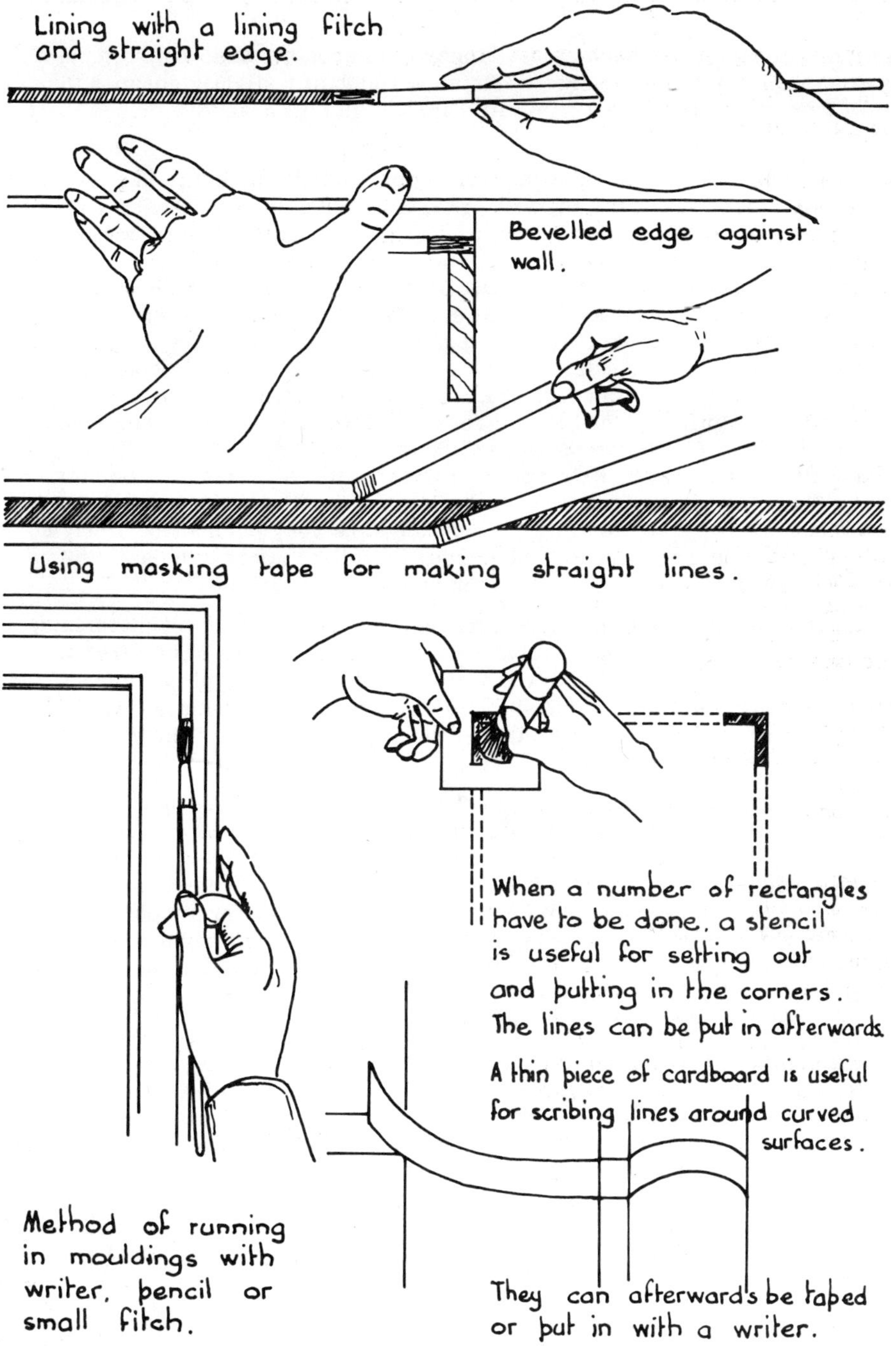

RUNNING LINES.

Water Mixed Material

Water mixed materials such as distempers, waterpaints and emulsion paints* are used to produce flat decorative finishes which are slightly cheaper than flat oil paints. All these materials are similar in appearance, although their properties are different.

Soft distempers are simple mixtures of pigments and glue size. The majority of soft distempers are white or broken white and are used mostly on ceilings or surfaces above head height (i.e. 1.5 m). The pigments used for making distemper are whiting and all those which are unaffected by alkali, such as OCHRES, ULTRAMARINE, RED OXIDE OF IRON, UMBERS, SIENNAS, ZINC YELLOW and BONE BLACK. For white, tints and very light colours, whiting is used as a base and is soaked in water two or three hours before the solution of glue size is added, the proportions being 3.5 kg whiting to 250 g concentrated glue size. For strong colours, dry pigments can be soaked in water in the same manner as the whiting, or pigments ground in water can be obtained. When thoroughly mixed, the distemper should be strained through a piece of coarse canvas. Ready-made soft distempers in thick paste form can also be obtained. Soft distempers reflect light very well and when used on surfaces which are not subject to hard wear and in dry conditions are very permanent. They cover well and usually only one coat is required. Soft distempers can easily be removed by repeatedly soaking with water and sponging off.

Waterpaints or washable distempers are made from a variety of materials, and, beside the pigments, may include resins, oils and waxes. They are a more permanent type of material and will withstand some degree of hard usage and washing. The white pigments used include zinc oxide, lithopone, as well as whiting. Waterpaints dry in two stages: (a) the evaporation of the water and setting of the size, (b) oxidation of the oil or polymerization of the resins.

Emulsion paints are made from synthetic resins e.g. polyvinyl acetate, alkyd or styrene dispersed in a water solution. The white pigments are chiefly titanium for high grade and cheaper pigments for less expensive materials. Drying is in two stages: (a) by evaporation of the water, (b) by polymerization of the resin. They dry quickly and can be second coated in two-three hours. All spots should be wiped up at once; otherwise they become difficult to remove, although a 1/5 acetic acid/water solution is usually effective. Emulsion paints should not be used in situations where hot steamy conditions occurs, since the particles of resin swell and contract and the coating ultimately flakes off.

Cement paints are mixtures of lime resistant pigments, white cement, waterproofers and adhesives. They are in powder form and mixed with water. Some manufacturers advise that the first mix should be left for 5-10 minutes and then a little more water added and the paint well stirred. If sand is added to make a fuller bodied paint the proportions should be 1 part sand to 4 parts cement paint. Do not mix more at a time than can be used within one hour. They can be used on cement rendering, asbestos, concrete and plastered surfaces, but are not suitable over woodwork or any decorative material; but most decorative materials can be safely applied over it.

* Formerly called 'plastic emulsions'

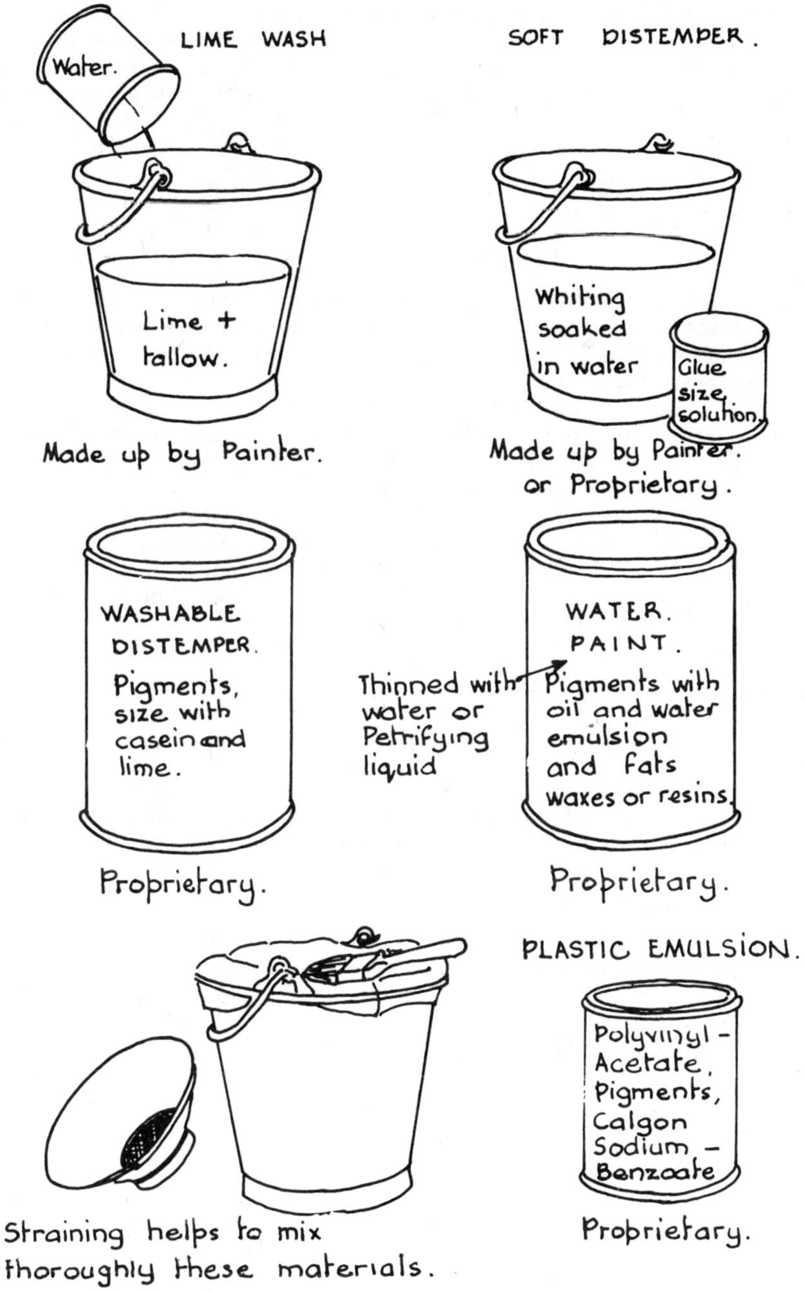

WATER MIXED MATERIALS.

Specifications for Application of Water Mixed Materials

All water mixed materials can be applied over surfaces which are clean and devoid of gloss. Soft distempers are mainly used on ceilings and walls. For new work, all that is required is a light rub with glasspaper and a careful dusting down. If two coats are specified, the first one should be in the form of a clearcolle, made up with $\frac{1}{3}$ distemper and $\frac{2}{3}$ glue size solution or cellulose adhesive. Surfaces previously treated with soft distemper should be well washed to remove the old material. Small stains on plaster can be coated with knotting, or a mixture of fine plaster and water before any distempering is done. Larger stains should be coated with flat oil paint. For very large stains or a severely cracked surface, it is better to cover with white lining paper.

Waterpaints and other hard types of material should never be applied over soft distemper. Waterpaints applied on new plaster should be thinned only with water, as this keeps the coating slightly porous and allows moisture in the plaster to escape. On new Keene's cement and surfaces which are 'hot', coating the surface with water before applying the waterpaint will greatly assist adhesion. The usual practice is to treat about 1 m^2 at a time, water and then waterpaint, until the surface is covered. On mature plaster, surfaces where hard usage is to be expected and on surfaces where resistance to weather is required, thinning should be done with petrifying liquid. The usual practice is to give a thin coat to assist penetration and obtain a grip, following with a thicker coat to finish. Surfaces which seem overloaded with waterpaint or which show signs of breakdown in the form of minute cracks or flaking should be stripped, but if cost will not permit, an oily sealer should be used, since this will help to bind the surface and act as a barrier to the water of the new material.

If emulsion paint is used, it is usual to mix equal parts of emulsion paint and water to give a 'mist coat' to seal the surface in much the same way as a 'clearcolle'. Emulsion paint can be applied by brush, spray or roller.

Cement paints should be mixed according to the manufacturer's directions. They should not be applied over damp materials containing an excess of sulphates such as new calcium sulphate plaster and some types of bricks; and for certain types of cement rendering special priming liquids are recommended. Cement paints are suitable for inside and outside work.

Silicate paints using a medium consisting of sodium silicate or other alkali silicates are useful to render inflammable materials such as timber, fibre board and fabrics less ignitable or to reduce the spread of flame. All the above materials are fire-retardant to a degree, providing thick coats are applied. Chlorinated rubber paints are useful for this purpose.

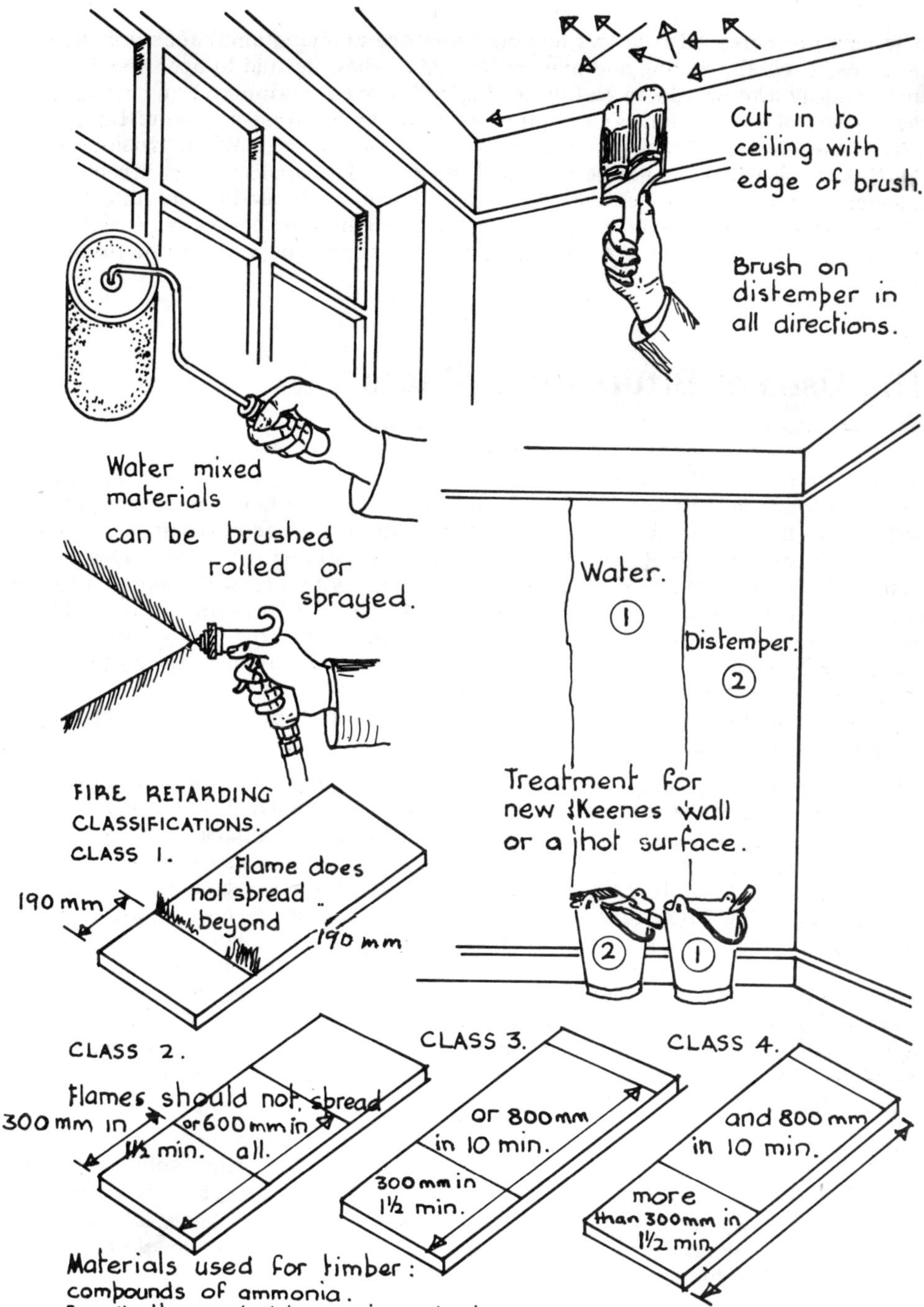

Materials used for timber: compounds of ammonia.
For all other materials: Sodium silicate, Ammonium phosphate, Calcium sulphate paints, Borax & Urea formaldehyde resin with ammonium phosphate.

METHOD OF APPLYING WATER MIXED MATERIALS.

Application of Emulsions and Distempers

It is usual to close all windows and doors before applying emulsions and distempers, to retard drying and prevent the joins showing; and to open them immediately afterwards, to hasten drying and prevent staining. For ceilings, the method of application is to commence near the greatest source of light, i.e. across the largest window, working in strips 0·5-1 m wide. When brushing, only the tips of the brush should be used and the materials applied in as direct a manner as possible without over brushing. Brushing should be done in all directions and generous coats applied. Roller application can be done with parallel strokes or in all directions. Spraying should follow the same broad principles.

The Uses of Bitumen Products

The best types of bitumen will withstand very strong sunlight without running or sagging and will remain elastic during very cold weather. Inferior types will not do this and tend to sag in very hot weather and become brittle in cold. Three types of bituminous products are used for building purposes: (a) waterproofing pastes, which are thick materials spread with a trowel used as a finish for concrete roofs, and repairing holes and cracks in valleys and gutters; (b) bitumen fibrous compounds, which are much thinner than the paste and are applied with a brush or broom for resurfacing surfaces already treated with bitumen, coating the inside of concrete storage tanks and the making of vertical and horizontal damp-proof courses; (c) bituminous coatings, which are applied by brushing, dipping or spraying.

Of the coatings there are four main types: (a) the standard black, which can be used for surfaces where the finishing colour is not important and is suitable for coating brickwork, metal, woodwork, concrete, and as a backing for stonework: (b) heat resistant blacks, which can be used for chimney stacks, stove pipes, radiators, and all metal surfaces which are subjected to temperatures of between 170°-200°C. At least 12 hours should elapse between the application of each coat and the heat should not be turned on before 36 hours; (c) aluminium bitumen coatings, which are useful where light needs to be reflected and the inside of the building kept cool; (d) coloured bitumen coatings, which have a limited colour range owing to the large asphalt content but are a great advantage in respect of colour to the standard blacks. Greens, dull red, browns and greys are the usual shades. They are useful for coating all types of building materials.

Bituminous materials are supplied ready for use, but should they require thinning solvent naphtha, xylene or benzene should be used, as white spirit tends to curdle them. On no account should these materials be heated, and they should be kept well away from naked lights and fires. If bituminous materials are being used in a confined space, as much ventilation as possible should be obtained, since the fumes given off are slightly toxic.

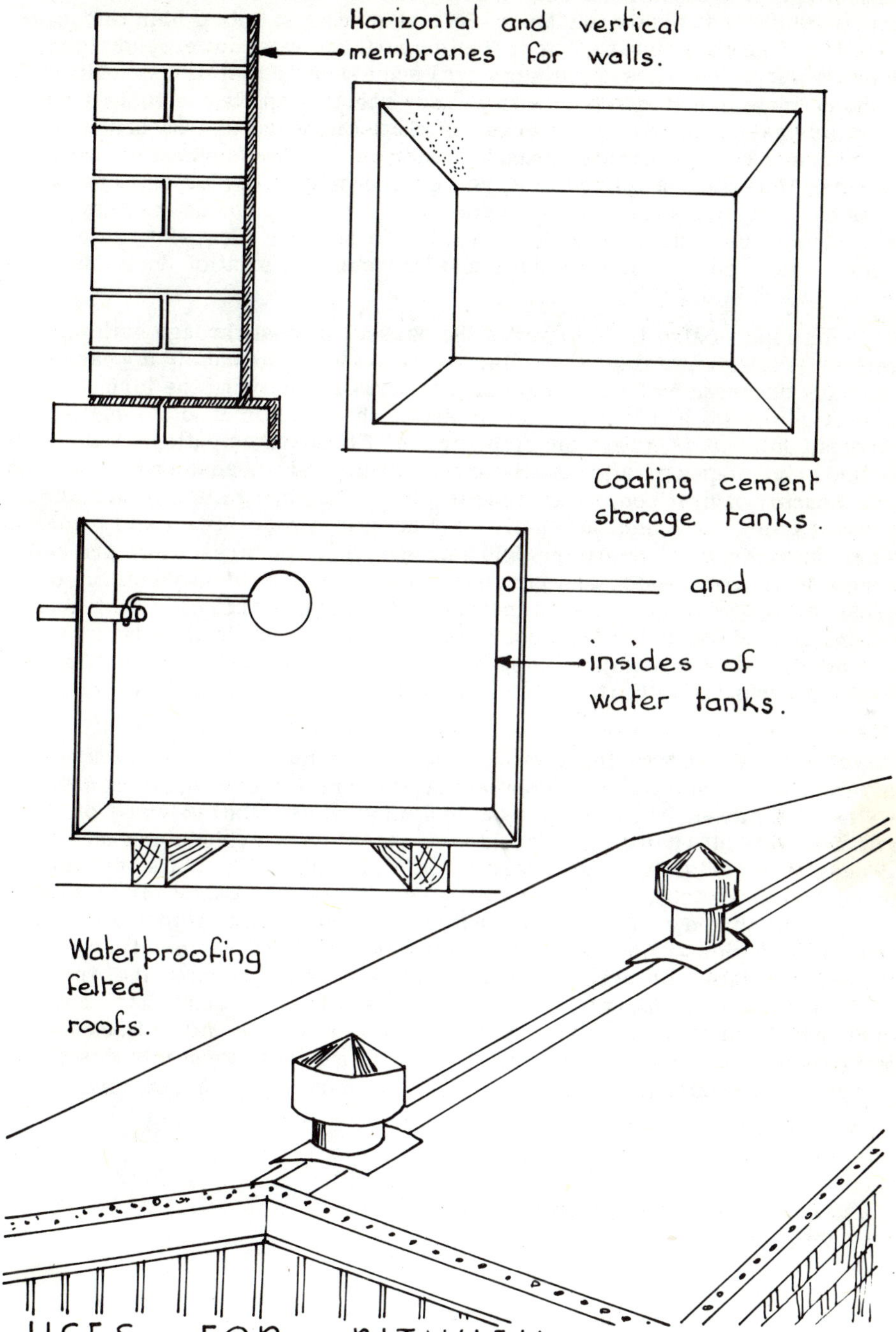

USES FOR BITUMEN.

Condensation and Pattern Staining

Condensation is caused by warm air striking cold surfaces, as when the hot tap in a bathroom or kitchen is turned on on a cold day and the warm vapour condenses on the cold tiles and windows. The air in a room with a high temperature will hold more moisture than if the temperature were lower, and if the temperature falls the excess moisture will soon become visible as condensation. Some materials conduct heat away very quickly and the inner surfaces in the room remain cold. Other materials insulate the heat and, in consequence, the surfaces inside the room are approximately the same temperature as the air inside. Materials which are porous, i.e. lime putty, soft plasters and building boards, wallpapers and textile fabrics, do not show condensation very quickly; but impervious materials such as metals, slate, glass, tiles, sirapite plasters and gloss paints do. Opening windows and improving ventilation does much to prevent condensation.

Pattern staining is also linked up with the passage of heat through building materials. Pattern staining is the discolouration which gradually appears on the surface decoration of walls and ceilings, usually following the form of the construction behind, and in time clearly shows the position of joists, rafters, girders, studding, etc. Where the structure is of timber, the pattern will assume light lines and shapes on a gradually darkening ground, because timber being a poor conductor of heat keeps the areas of plaster adjoining slightly warmer than the spaces in between. When air circulates in a room it tends to be deflected from surfaces which are warm and strike against surfaces which are cold, and the minute particles of dust carried by the air are gradually built up on the cold or colder surfaces. With metal construction, the effect is dark lines on a light ground, because metal being a good conductor of heat withdraws the heat from the materials adjoining so that these areas become colder than the rest of the surface. With pattern staining it is always the cold areas which become dark.

To overcome pattern staining and condensation and make a permanent cure, it is necessary to insulate the surface so that the temperature will be constant over the whole area. This can be done by lining with insulating board, or coating with a material having good insulating properties, provided a sufficient thickness of material (approximately 2 mm) is applied. Materials such as 'Seculate' anti-condensation compound or cork paints are useful for this purpose and, if required, can be finished with distemper or flat oil paint. On no account should gloss paints be used as finishes over these materials, otherwise their insulating properties will be destroyed. Another method of treatment is to double line with lining paper. For ceilings with voids above, laying metal-faced paper or fibre glass quilt over the joists is effective and on flat concrete roofs a layer of shingle 75 mm deep. Vermiculite laid between joists is also a good non-rigid insulating material which can be used to fill up awkward corners and voids.

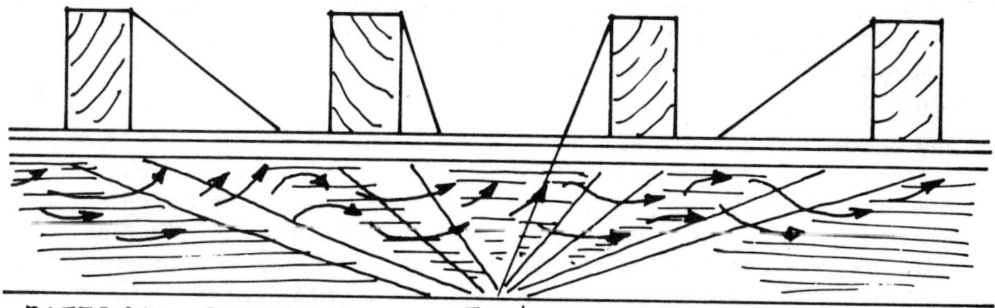

PATTERN STAINING Timber Construction.

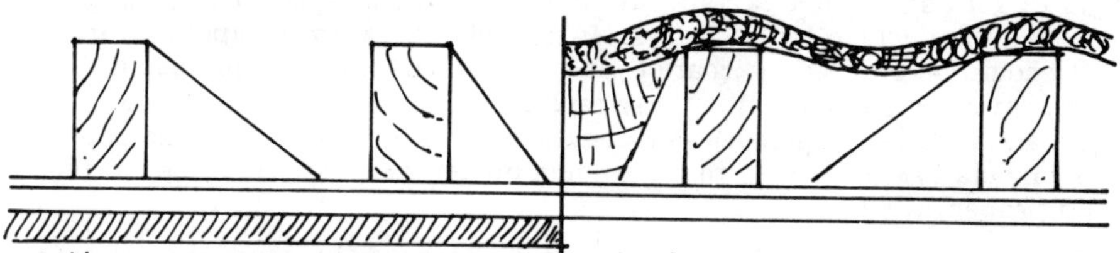

Anti condensation compound at least 2mm thick. Fibre glass quilt.

TWO METHODS OF COMBATING PATTERN STAINING.

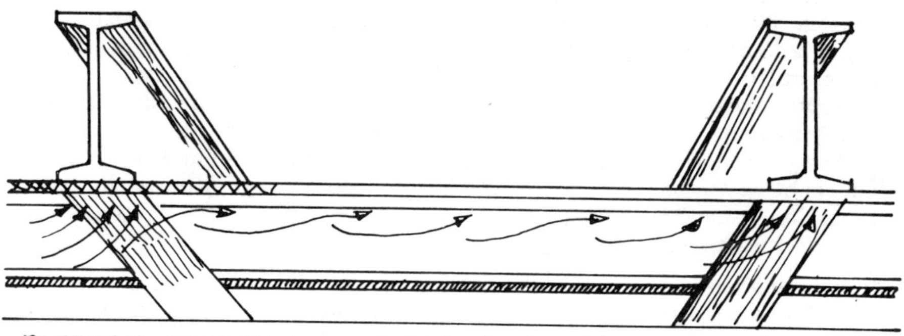

PATTERN STAINING Rolled steel joist and angle iron.

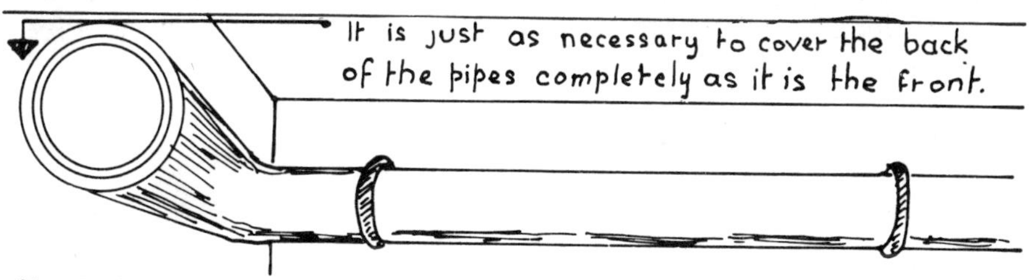

PATTERN STAINING AND CONDENSATION.

Paperhanging

PREPARING THE SURFACE

Surfaces for wallpapering must be carefully prepared and well rubbed down to provide a perfectly smooth face. Different types of surface need special treatments and where an extra thick wallpaper is to be hung the surface should be hung with lining paper first.

New lime plaster should be allowed to dry out properly, then glass papered to remove all nibs and efflorescence. Dusting off and the application of a coat of weak glue size complete the preparation. For first class work, a coat of alkali-resistant primer is recommended before the application of the glue size. New hard wall plasters need rubbing down, dusting off, and a coat of glue size. New Keene's and Parian cement should be coated with a ready mixed Keene's cement primer thinned with white spirit or a mixture of 1 part white lead, 1 part japan gold size and 4 parts white spirit by volume, applied 'following the trowel' and the surface left 2-3 days. If the primer cannot be applied within this time it should be omitted. A coat of glue size should then be applied. Cold closely-trowelled surfaces like these tend to encourage moisture to condense on them and it is recommended that they are 'crossed lined' (a lining paper hung horizontally) or hung so that the joins do not coincide with those of the finishing paper.

Surfaces already papered should be stripped. This can be done by repeatedly wetting the surface to soften the paper, enabling it to be removed with a broad bladed stripper. After stripping, the surface should be washed down to remove old paste, size and particles of paper; all holes, cracks and other defects made good, and allowed to dry. Glass papering and a coat of glue size complete the preparation. Where several thicknesses of paper have to be removed, use can be made of a Sanderson wallpaper stripping machine. Where wallpaper has to be hung on walls which are already papered the joints of the old paper should be carefully rubbed down and any loose edges stuck back or torn off. Sizing completes the preparation. To remove varnished paper, a mixture of caustic soda and paste can be brushed on with a fibre brush to soften and destroy the varnish, enabling the paper to be removed in the usual way. Afterwards the surface should be washed with water and vinegar to neutralize the caustic. Paint solvents are also useful.

Soft distempers need to be thoroughly removed by repeated soakings in water; washable distempers should be rubbed down dry and primed with an oily primer prior to sizing. Painted surfaces should be washed down with sugar soap or soda to kill the gloss, rubbing with waterproof glass paper or pumice block during the process to provide 'key'. After rinsing with vinegar water, apply one coat of glue size mixed with a handful of plaster or whiting to provide a slight degree of absorbency, and cross line allowing the paper to soak well before hanging.

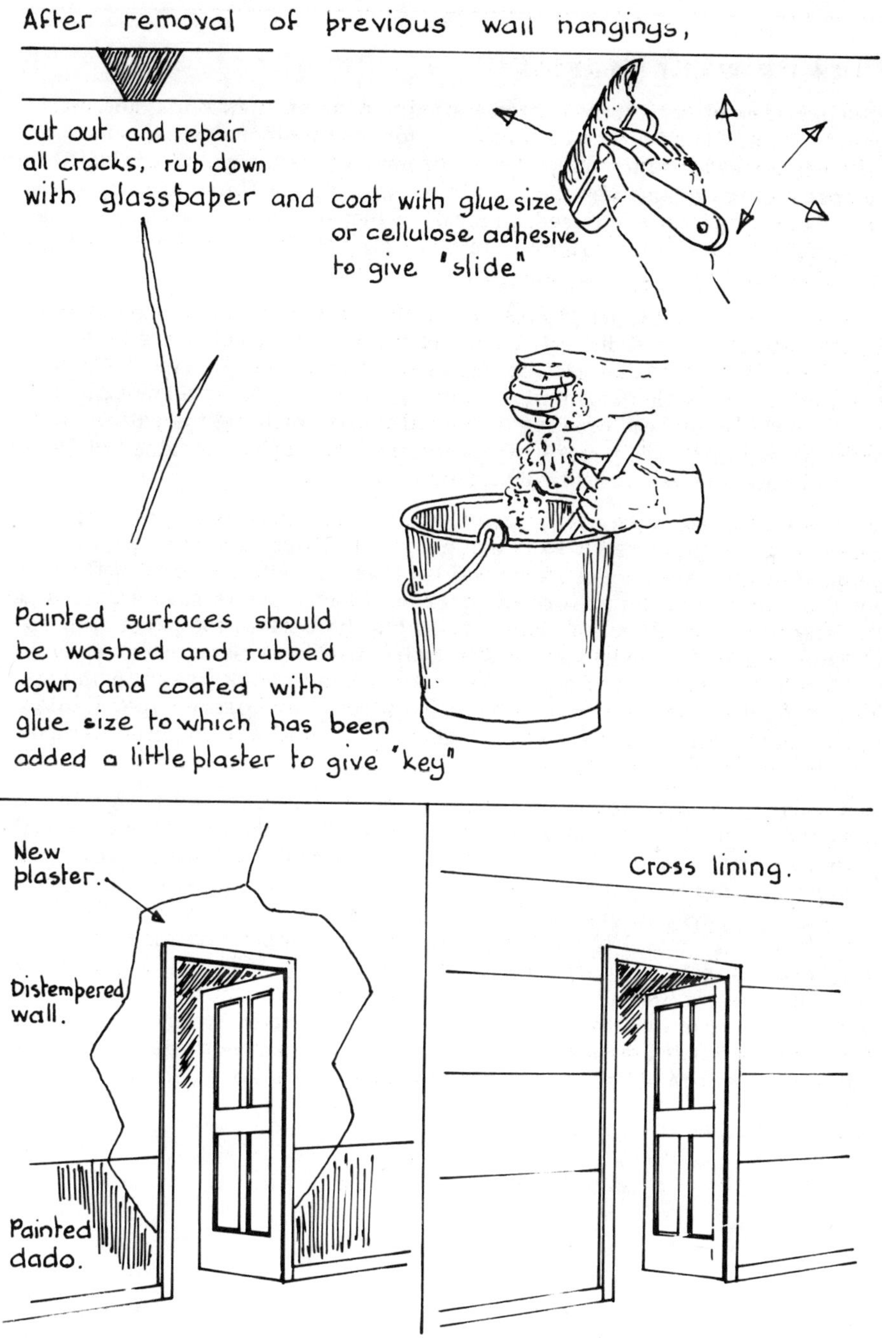

SUITABLE PREPARATION FOR THE HANGING OF A THICK WALLPAPER.

PREPARATION OF SURFACE FOR PAPERHANGING.

Paperhanging

TYPES OF WALLPAPERS

Wallpapers are of two types: (a) foundation papers such as lining and pitch paper, which are used either as a backing for more expensive papers or to strengthen a weak surface and act as a ground for painting materials; (b) finishing papers, which can be obtained in almost any type of finish and any colour. Lining papers are self-coloured, generally white or brown, but can also be obtained in light tints. Pitch paper is brown lining coated on one side with bitumen and is used on walls that are damp.

The main types of finishing papers are: Pulps, which are the cheapest and have the pattern printed directly on the paper; Grounds, which are better quality papers grounded or coated with colour before the design is printed; Mica prints, which are papers with a satin-like sheen produced either by adding mica to the ground colour before the printing of the pattern or printing the pattern itself in mica; Moirés, which are papers with a watered-silk effect, obtained either by a fine delicate emboss or by ordinary printing.

Textures imitating leather, wood grain, plaster and textiles are obtained by embossing the paper after it has been coloured. More expensive papers of this type have two thicknesses of paper welded together with the back of the paper having the texture of the pattern in reverse. Plastic prints have a raised pattern resembling the effect of plastic paint but the back of the paper is perfectly flat. Other papers include wax finished, varnished, and 'sanitaries' which are printed in oil colour. More expensive materials are flock papers, which have a raised pattern similar to cut velvet; a high relief embossed paper (hollow backed) sold under the trade name of 'Anaglypta', and a solid relief compound called 'Lincrustra Walton.'

The standard length for machine printed papers is 10·05 m, with a variation not greater than 1½ per cent; and a width of printing 533 mm, with a variation not greater than 5 per cent, giving a superficial area of 5·6 m^2 approximately per piece. Borders in rolls are 8-11 m in length, with a variation of 1½ per cent, and when supplied in coils 11 m or multiples of 11 m. Stiling borders are generally 127 mm, 177 mm or 266 mm wide. French and American papers are different in size. French papers are 8 m long by 457 mm wide, giving a superficial area of 3·6 m^2. American papers are 7 m long by 457 mm wide; 'double' rolls 14 m long by 711 mm wide are also obtainable. When selecting and hanging patterned wallpapers as opposed to stripes, dots and powderings, attention should be paid as to whether the pattern is a horizontal or drop repeat.

DROP PATTERN. WIDTH IN THIS CASE SHOWS ONE AND A HALF REPEATS.

HORIZONTAL PATTERN. WITH SMALLER UNITS GOING RIGHT ACROSS THE FULL WIDTH.

TYPES OF WALLPAPER DESIGNS.

ADHESIVES AND TRIMMING

The adhesives used for paperhanging are ready-made powder pastes, cellulose pastes, dextrine, and flour or starch which are made up by the paperhanger. Some ready-made pastes can be mixed with cold water, which is an advantage on jobs where it is difficult to obtain boiling water or heating facilities. The method of mixing is to put 5 litres of water in a clean bucket and slowly add the paste, stirring briskly all the time. Cellulose pastes are mixed in exactly the same manner. After 15 min the mixture swells and thickens considerably and requires thinning and re-stirring before use. For lightweight papers extra water can be added. With cellulose pastes the mixture can be diluted and used as a 'size' for preparing the walls, thus doing away with the necessity for using another material and bucket. For high class papers tub paste should be used.

Flour paste is made by putting 1·75 kg of flour in a bucket and mixing with water until a smooth, slack batter is obtained. 5 litres of boiling water are then slowly poured on the batter, stirring all the time until the paste thickens. Starch paste is made in a similar manner, 250 g of ordinary starch being used instead of the flour. Dextrine is a paste of great strength. For heavy papers such as Lincrusta, it is usual to use equal parts of flour paste and dextrine, but for fixing thick strappings and borders it should be used neat. Rubber fixatives are also useful for this purpose. Ready-made adhesives keep better than flour or starch pastes and it is essential to use paste that is fresh and sweet. Paste of any type should be strained before use.

Before wallpaper can be hung, the selvedges or outer edges which protect the pattern must be removed. This can be done by machine at the merchants for a few pence per roll, but the best way is for the craftsman to trim it just before the paper is hung. Shears and pocket-trimmers like the Morgan-Lee are used for most general types of paper before they are pasted. When using shears, the easiest way is to sit with legs outstretched in front with feet 0·3 m apart. The paper is cradled at the feet, trimmed with the shears held in the right hand, and pulled and re-rolled with the left. With pocket-trimmers, the paper is laid on the table, the distance between the cutting wheels and the back of the guide adjusted, and the guide introduced to the edge of the paper and the trimmer drawn along. For thicker papers which require an undercut edge to avoid the thickness of the paper showing at the join, it is usual to paste the paper first, fold it, and then trim with a knife and straight edge, using a strip of sheet metal underneath. Use can also be made of the Ridgley wheel trimmer, a plane-like instrument which is fixed on to a special grooved steel straight edge, along which it travels.

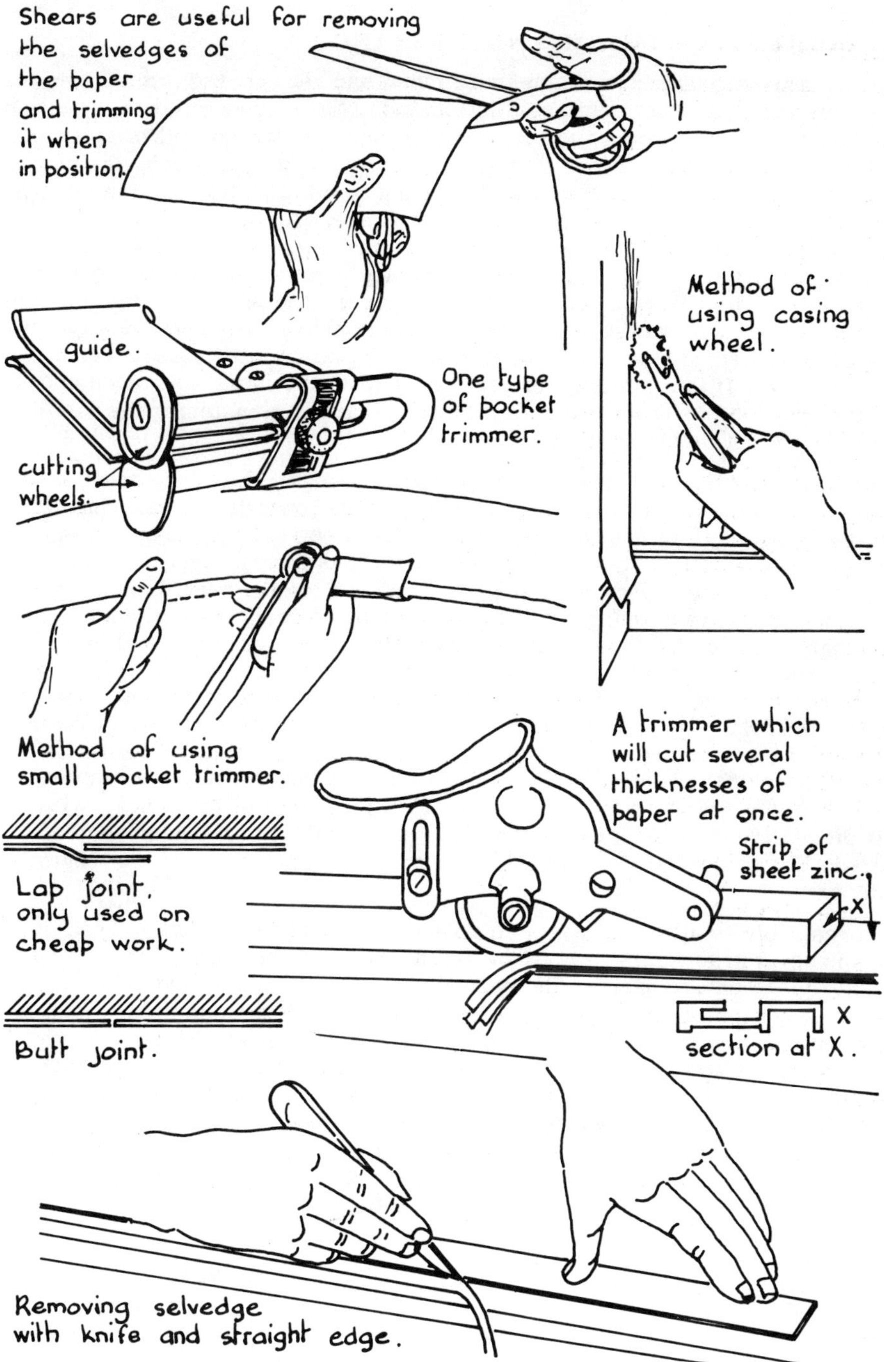

TRIMMING OF PAPER.

MEASURING, CUTTING AND PASTING

The art of paperhanging is to 'measure twice and cut once'. In this way mistakes are avoided. The most reliable method of finding the number of rolls of paper required for a room is to walk round the room with a roll of paper, mark off the number of long lengths that are required, and divide this by the number of lengths that can be cut from a roll, bearing in mind that allowance must be made for scribing, waste and type of paper. Thus, with plain lining paper, it is essential to allow 75-100 mm extra on the actual measurement for scribing at the top and bottom edges. With large patterned papers it is necessary to choose some part of the pattern to form the line along the top, and it is from this point that the lengths are measured. There is nearly always a surplus of paper between the measured length and the next chosen point and it is usual to cut between the two. If this distance is very great (300-450 mm), it is more economical to cut lengths from two rolls at a time, carefully matching the joining of the pattern first before cutting.

When the number of full lengths have been cut, they are laid face down on the bench over to the far side, with the matching edge towards the paperhanger, and the first length is drawn level with the front edge of the bench. In this way the top of the bench is completely covered and paste cannot get on it and so to the face of the paper. It is then pasted, working from the centre of the paper out to each edge, care being taken to keep the paste from the face side. When the length of paper has been pasted, the end is folded over and gently pulled along the bench so that a further section is in position for pasting. When the end is reached, the end is folded as before in the opposite direction. For ceiling papers, the folds are made about 200-250 mm wide concertina-wise. Where narrow lengths are required, as at corners, paste and fold a full length as before, measure the required width plus 10 mm to allow for any unevenness and mark the line by using a rule as a guide along the edge of the bench. With the last piece the far edge is pasted, then the paper pulled to the near edge and pasted. The paper is then folded and the process repeated until the length is finished.

It is desirable to allow the paper to soak before hanging, to obtain equal stretch, and the usual practice is to start by pasting two lengths, hanging one, and pasting another piece before hanging the second, and so on. Fragile and delicate papers should be hung more quickly without much soaking. Stout and extra thick papers may need pasting twice. Paste two lengths as before, paste the first one again, and hang straight away. Papers coated with an adhesive are also obtainable. They are supplied with a 'throw away' trough, which filled with water is placed at the bottom of the wall where the paper is to be hung. The cut length of paper is passed through the paper trough and straight on to the wall being brushed and trimmed afterwards in the usual way.

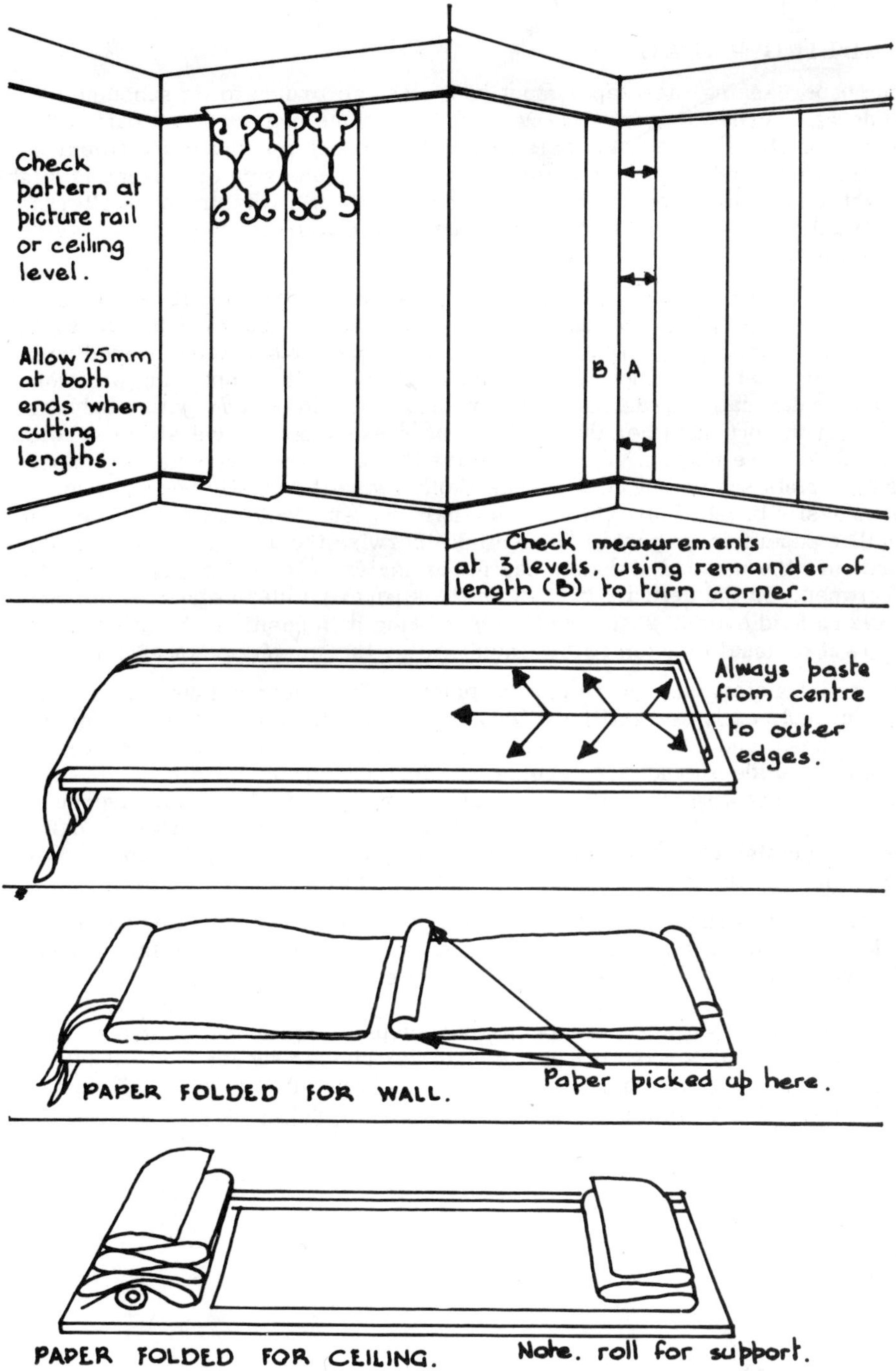

MEASURING, CUTTING & PASTING PAPER.

Paperhanging

HANDLING PAPERS

When paperhanging each paper must be treated according to its substance and design. Although great care is taken, variations in colour in a batch of papers sometimes occur, and it is advisable to check for shade and colour so that any variations can be grouped together. Often one side of a paper is slightly lighter than the other and if the pattern will allow, reversing each alternate length will bring the edges of the same depth of colour together and so avoid a striped effect.

Thin papers such as pulps should be trimmed well in, pasted with a round (that is medium thick) paste and hung immediately. Satins should not be oversoaked and require smoothing with a perfectly clean brush or soft felt-covered roller. Embossed papers must not be oversoaked otherwise the embossing will work out and make matching difficult; the smoothing should be done with a dabbing motion of the brush rather than in hard sweeps. On no account should a roller be used on these papers. Ingrains require the same treatment as pulp papers. Plastic prints should not be oversoaked otherwise the plastic becomes soft and is easily brushed off. On moirés a hard roller should never be used. On metallic papers fresh paste is essential otherwise the acid in the sour paste discolours the metal, and double pasting is advised with a few minutes interval in between. On the delicate surface of flock papers a clean piece of thin paper should be held over it when brushing or rolling it in position. For ordinary papers it is usual to seam roll the joints after the completion of each flank.

High relief materials require careful planning, trimming and cutting as mistakes cannot easily be rectified. They require a thick paste, such as a mixture of flour paste and dextrine or flour paste and glue. 'Anaglypta' should be pasted, put aside to soak and become pliable, then repasted and hung. It should not be folded at a sharp angle and brushing should be firm but not vigorous. It is best positioned with a soft brush and the flat areas pressed with a damp cloth. Push pins can be used to secure it temporily and when dry open joints and pin holes can be made good with plaster and washable distemper.

Lincrusta needs sponging on the back with hot water several times and laid aside to soak until it will absorb no more water. Care should be taken not to bend it at too sharp an angle since it breaks easily. It is usual to soak two pieces, repaste the first and hang it while a third piece is soaking and so on. Removal of surplus paste with a damp cloth completes the operation. Superimposed borders are not soaked but pasted with dextrine or rubber paste at full strength applied with a palette or broad knife. The adhesive should become tacky before the border is rolled into position. Do not apply glue size over lincrusta.

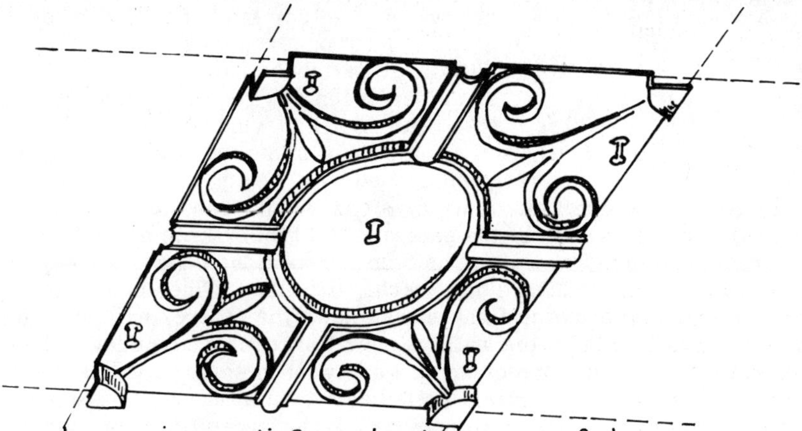

· when using relief materials careful setting out is necessary.

· when in position anaglypta can be secured temporarily with push pins.

wetting the back of lincrusta.

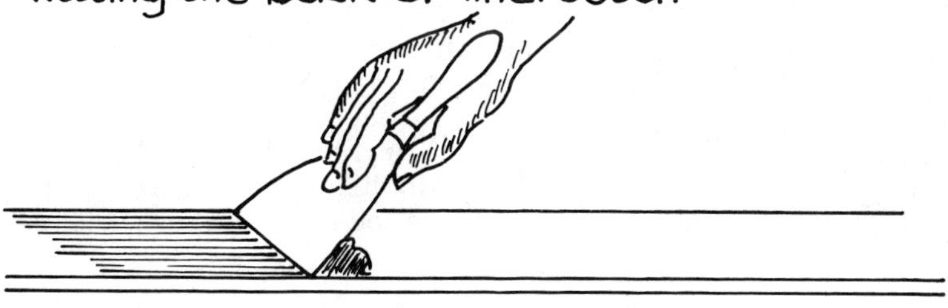

applying adhesive to a lincrusta border.

TECHNIQUES WITH RELIEF MATERIALS.

THE FINAL STAGE

When all paintwork is dry, start papering from the greatest source of light, i.e. ceilings across the largest window, walls on either side of a window, or other prominent feature. A large patterned ceiling paper, however, has to be worked from the centre of the ceiling. With large patterned papers it is usual to centre the chimney breast, mark off the number of lengths required to reach each side of door or window and at the marks near door and window to plumb lines to ensure the paper is hung vertically. On a pair of steps the paper is held in position slightly away from the wall, and opened by grasping the top corners and letting the folds drop down by the paper's own weight. Any jerk should be avoided by taking the weight of the paper on the foot or knee and slightly lowering the hands. With one top corner held on the wall, the paper can be pivoted into position by slightly raising or lowering the other. When positioned correctly, the smoothing brush is first passed down the centre, then left and right until the paper is firmly on the wall. A pencil should then be drawn along the top angle between ceiling or picture rail and wall to scribe a line for cutting. The top edge of the paper is gently pulled away, cut along the line and brushed back again. This is repeated at the bottom. When cross lining or hanging paper horizontally, a simple scaffold of two pairs of steps and a board of adequate length is required. Start on the right, open the first fold with the right hand, and place the edge of paper at the correct level; press the end of the paper into the corner and brush the paper smooth continuing to open the folds and brushing until the other end is reached. The ends are trimmed as before. For low ceiling work, a simple scaffold of two pairs of steps and a board usually suffices; for higher ceilings, a movable staging or board supported by ledgers will be required. The folded paper is supported by a roll of paper held in the left hand, leaving the right free to manipulate paper and brush. The top fold is fitting to the line marked on the ceiling and the paper brushed on, keeping the left hand held well up, moving along the board and following closely with the brush as the folds are opened. With a ceiling rose or similar obstruction, radial cuts are made and each section scribed around the edge. Joints and angles will need careful pressing with a steel or boxwood roller after each two or three lengths have been hung. Paste which gets on the paintwork should be sponged off immediately, otherwise it may cause it to crack.

For trimming around a mantelpiece with a top of more than one level, position the top of the paper but keep the lower half folded until the top level is reached. Scribe along this and make the necessary cut. Then cut to within 10 mm of the vertical edges until the second levels are reached on either side and scribe and cut as before. Scribe the paper into the angle of the mantelpiece and along the top of the skirting and cut both edges. Holding the lower half of the paper from the wall make the final trimming at the top and vertical edges and brush into position.

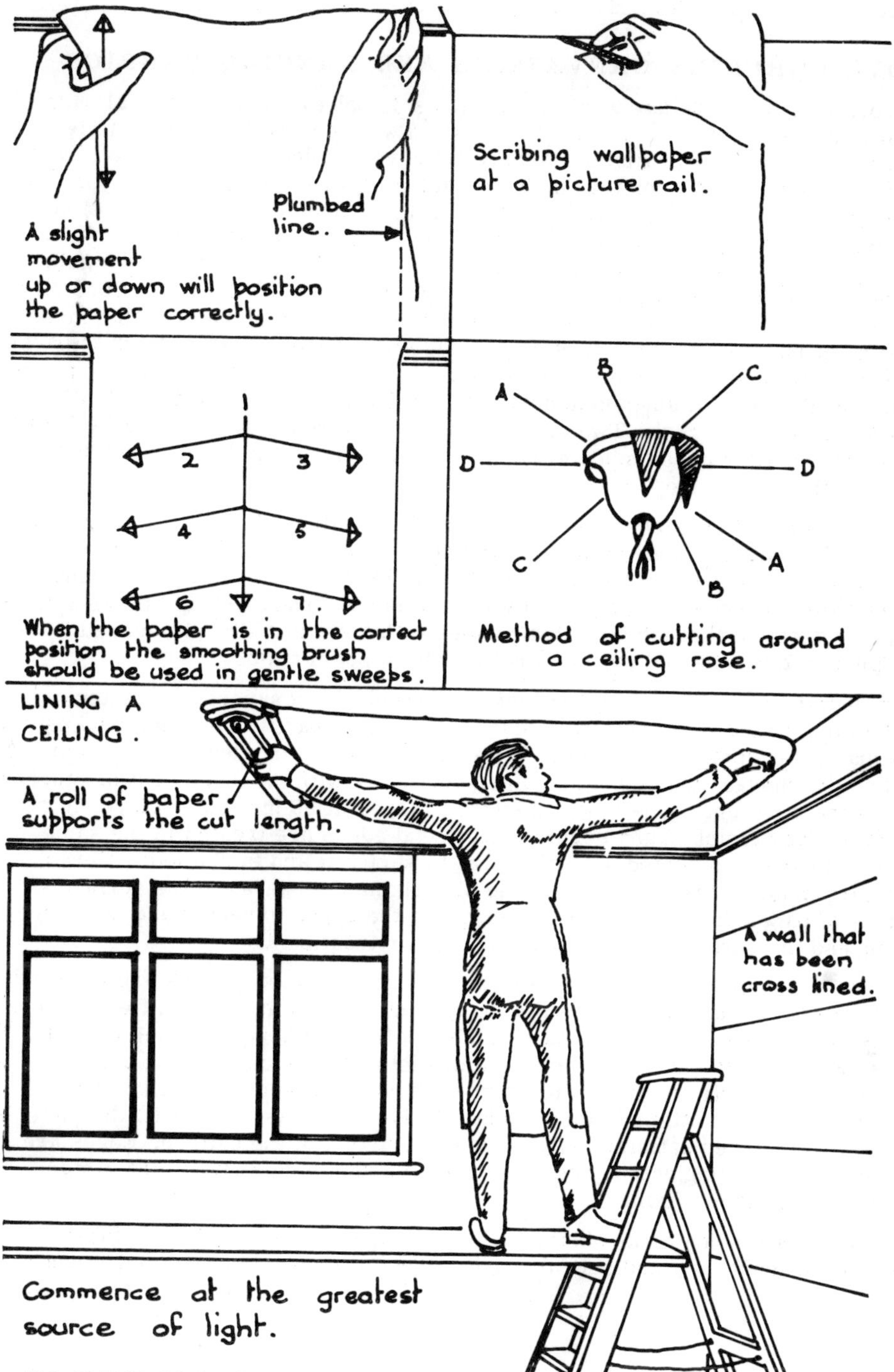

PAPERHANGING — THE FINAL STAGE.

COVE CORNICES, CANVASSING AND HANGING FABRICS

Decorative coves are useful where cornices do not exist, or to conceal wide cracks at the junction of ceiling and walls. Two types, both approximately 125 mm in girth are available; 'Anaglypta' cove cornice supplied in rolls and fixed in the usual way for that material, and a paper encased plaster type coving like 'Gyproc' available in lengths of 3, 3·6 and 4·25 m, which is either stuck, nailed or screwed. To fix by adhesive: (1) Ensure all surfaces are clean, free of loose material, soft distemper and wallpaper. (2) Set out guide lines 85 mm from the angle on both ceiling and walls. (3) Cut mitres and mix adhesive by slowly stirring 2 kg of adhesive into 1 litre of water which makes sufficient material to fix 8 m of coving. (4) Apply adhesive 3 mm thick to backs of both ceiling and wall members along their entire length and push into position. (5) Remove all surplus adhesive with a knife or trowel and finish all angles with a wet brush. Two operatives are required to handle lengths 1·5 m or over. Smaller coves made of expanded polystyrene can be obtained and are fixed with an adhesive.

Coarse fabrics like backed canvas or burlap are attached direct to the prepared walls. Trimming is done with knife and straight edge. Stiff paste is used and the material soaked well before being hung. Unprepared hessians are hung dry being well brushed onto the pasted wall. Joins not covered should be well lapped; the edges cut through both thicknesses with knife and straight edge, the surplus pulled away and the joins rolled with a seam roller.

Although the canvassing of partitions has largely been superseded by the use of building board it is used on grooved and tongued partitioning as a backing for wallpaper or an expensive fabric. The surface should be cleaned down thoroughly and the canvas cut in lengths, joined vertically with the seams sewn. Starting at the top right corner the top edge is tacked at 50 mm intervals, then the bottom edge stretched and secured. Both end edges are treated in the same way: the canvas is coated with glue size to shrink it and fasten loose fibres, but care must be taken to prevent the size fixing the canvas to the surface. Strips of paper 75 mm wide are stuck along the edges to neaten and reinforce them, and the surface lined before any further treatment.

Expensive fabrics are hung as follows: 50 × 25 mm wood battens are fastened to the walls top and bottom at suitable 1 m intervals. Any inequalities between walls and battens are packed out to enable nailing, without loosening the battens. Fine quality canvas, domett or flannelette is stretched on the battens. The fabric is centred, lightly tacked at each corner and at 50 mm intervals along all sides, no attempt being made to drive the tacks home until the material has been stretched satisfactorily without distortion of the pattern. Edges are finished by covering with a wooden decorative moulding or furniture gimp, using gimp pins for fixing.

With all types of material whether it is wallpaper, coving, canvas or fabric, careful measurement and setting out is essential before the material is cut.

FIXING PLASTER COVING.

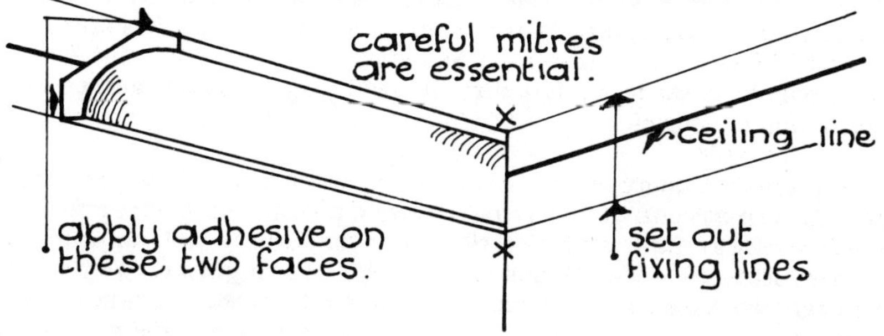

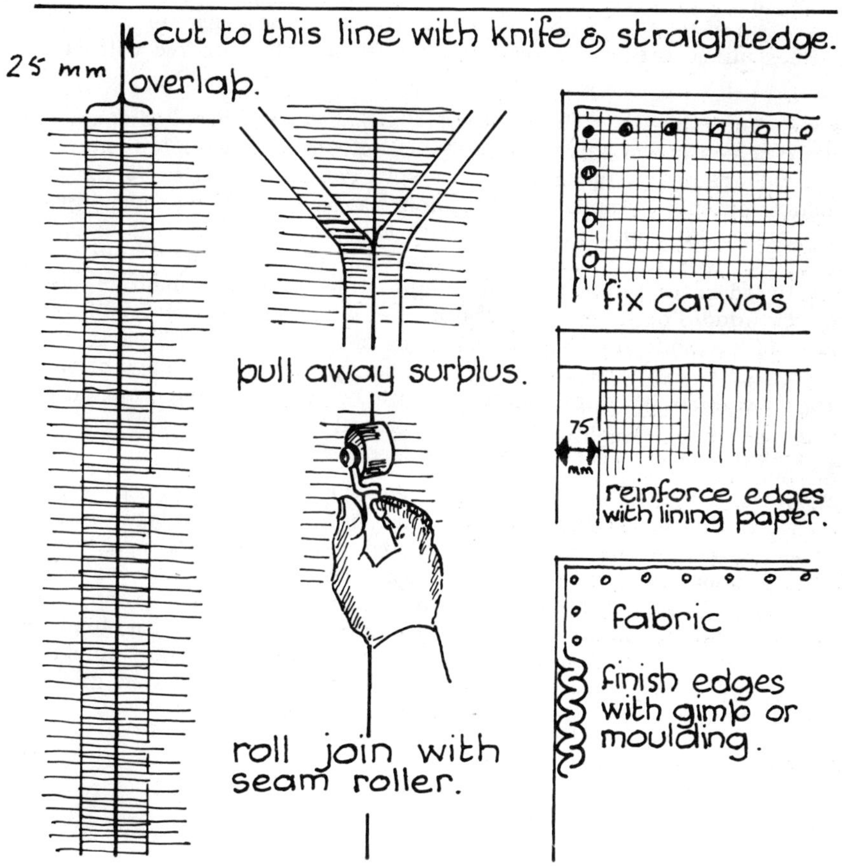

HANGING HESSIAN AND FABRICS.

Colour I

If a ray of sunlight is shone through a prism, it can be broken up into a continuous band of colour, changing gradually from red at one end to violet at the other. For our purposes we can assume that white light is made up of six separate colours— red, orange, yellow, green, blue and violet. When white light falls upon a surface some of the colours making up the light are absorbed and some reflected. Our sensation of colour depends on which rays are absorbed and which reflected. A white object reflects all colours equally, but a red object will absorb all colours falling on it except red, which is reflected. Thus we perceive the object as being red. Similarly, a green object reflects only the green component of white light, other colours being absorbed, and so we see the object as being green.

Pigments cannot be obtained so pure as light, and if red, orange, yellow, green, blue and violet pigments of spectrum strength are mixed together the result is dark grey or black. Basically, a six colour theory works well, as it gives each colour an equal value and includes emerald green and purple, two colours which are often termed secondary colours but cannot be mixed. The six colours at full strength - red, orange, yellow, green, blue and purple - are simple hues. When any two of these colours are mixed together, they become intermediate hues. Hues mixed with white are termed tints, and those mixed with black, shades.

If the six simple hues and their intermediate hues are arranged equidistant round a circle, any two colours exactly opposite will be complementary, and any two colours nearly opposite will be split complementaries. Pairs of complementary or split complementary colours mixed together produce grayed or broken colours. If a circle is drawn with 12 equal sectors and is divided by means of 5 concentric circles in to a small circle with 5 outer rings, the colours can be plotted in the following manner: the simple and intermediate hues will occupy the middle or third ring. The 2 outer rings can be filled with (2) equal mixtures of hue and white and (1) mixtures of 1 part hue and 2 parts white, giving tints of two different intensities. The fourth ring will have broken colours obtained by mixing 2 parts hue with 1 part of its conplementary colour opposite. The fifth and inner ring will be filled with shades obtained by mixing equal parts of hue and black. The inner circle will be black.

The most important pigmentary colours are the reds, yellows and blues because they cannot be obtained by admixture, but by using two types of red (vermilion and crimson), two yellows (lemon and cadmium yellow), and two blues (cobalt and monastral) with white and black many other colours can be obtained. Hence, reds, yellows and blues are termed primary colours, and the oranges, greens and purples obtained by mixing red and yellow, yellow and blue, and blue and red respectively, are known as secondary colours. Tertiary colours can be obtained by mixing two secondary colours together, orange and green making citrine, green and purple slate, and purple and orange russet.

THE SOURCE OF ALL COLOUR.

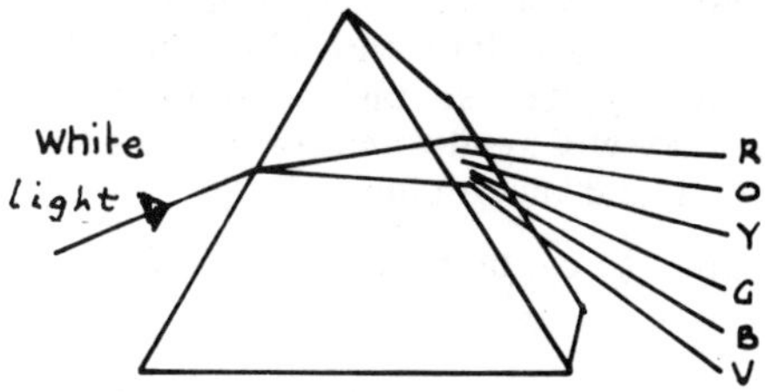

ABSORBED & REFLECTED RAYS.

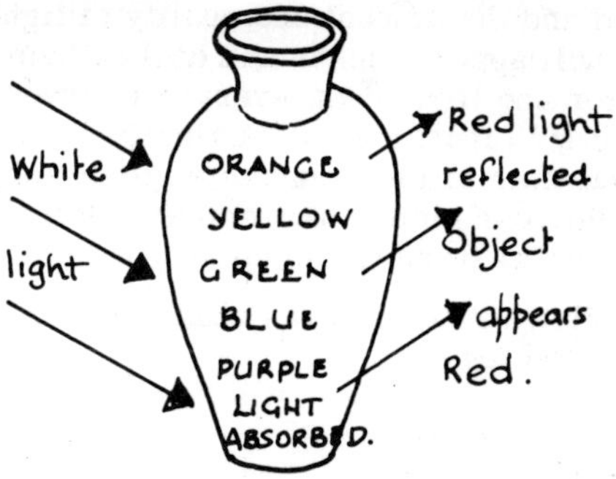

A. & B. TINTS. = SIMPLE & INTERMEDIATE HUES + WHITE.

C. = HUES.

D. BROKEN OR GRAYED COLOURS = MADE FROM HUES WITH THEIR COMPLEMENTARY HUE.

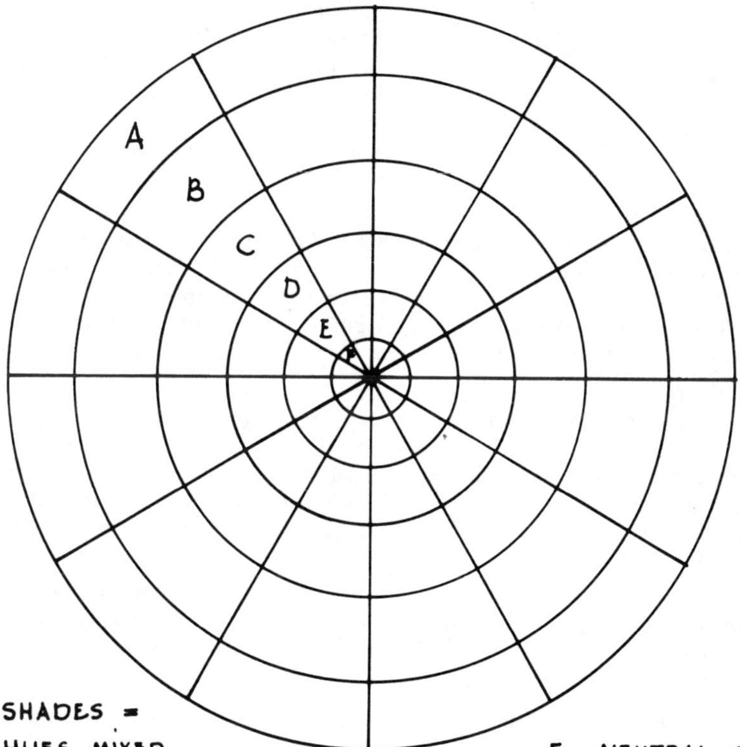

SHADES =
E. HUES MIXED WITH BLACK.

F. NEUTRAL GREY. OR BLACK.

COLOUR I.

Colour II

The appearance of any colour is influenced by the colour next to it, its background and the amount and quality of light falling upon it. A standard hue on white will appear richer and darker than when on light grey, and very much brighter and lighter on a very dark grey or black. A blue strip on an orange ground will appear more distinctly blue than the same colour on a blue ground. Neutral tints in the immediate vicinity of strong colours assume to some extent their complementary colour. Green with grey will tend to cause the grey to appear warmer (reddish) and the green colder and more blue.

The following table shows what appears to happen to certain hues when they are placed together:

Hue inclined to	Hues placed together		Hue inclined to
Violet	Red	Orange	Yellow
Violet	Red	Yellow	Green
Orange	Red	Blue	Green
Orange	Red	Violet	Blue
Red	Orange	Yellow	Green
Red	Orange	Green	Blue
Yellow	Orange	Violet	Blue
Orange	Violet	Green	Blue
Orange	Yellow	Blue	Violet
Yellow	Green	Blue	Violet
Yellow	Green	Violet	Red
Green	Blue	Violet	Red

Any object of a definite hue will look brighter if illuminated by white light or light rays of the same hue; but if the colour of the light is changed from the same hue as the object to that of its complementary, the object will appear black or dark grey. Thus a red object will reflect red light in white or red light, but if the light changes to blue-green, it will appear black.

All types of artificial lighting change or alter colours and efforts to correct these changes have resulted in so called 'daylight' lamps. Incandescent light absorbs blue-green and colours look warmer than in normal daylight. Fluorescent light absorbs red-orange and causes colours to look colder. Sodium light absorbs red-blue-green and red, and makes most colours look either yellow or black. Mercury light absorbs red-yellow and colours in this light look bluish and colder.

Certain pairs of colour combinations show up more clearly than others. Black on yellow, green on white, red on white, blue on white and black on white are good examples, but if the order is reversed the sharpness or contrast is not so apparent, effects which are well known to signwriters.

Combinations of very similar hues can be disturbing. A pure red with a bluish red will tend to change the red to orange-red and appear unstable.

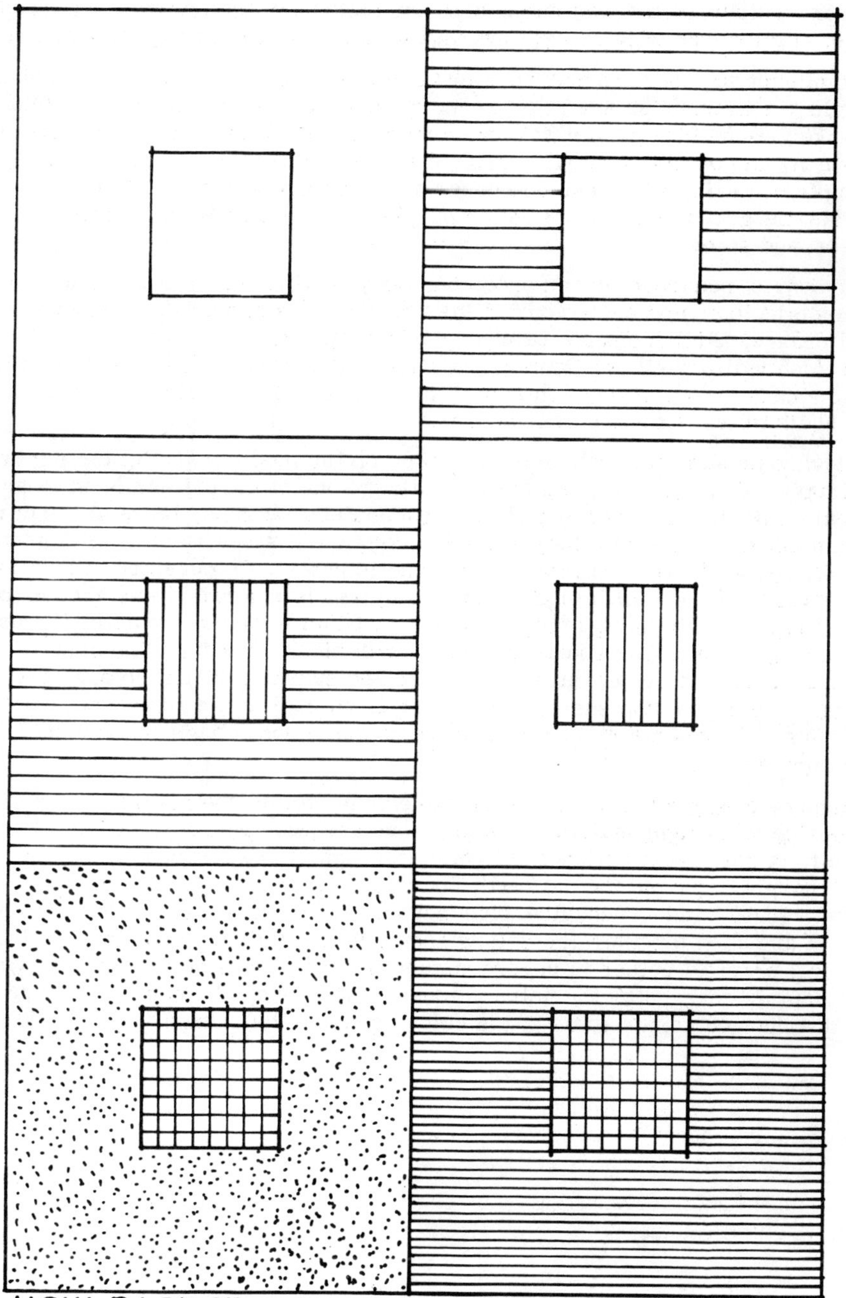

HOW BACKGROUNDS INFLUENCE LEGIBILITY.

Colour III

There are many types of colour harmony: monochromatic, when use is made of three or more tones of the same colour, i.e. pale, light, middle and dark blue; dominant, which is similar to monochrome, but with the use of small areas of another colour, so that the one colour dominates, i.e. light, middle and deep blue, with touches of lemon yellow. In this case the yellow would be an accent.

Complementary colours are when two colours exactly opposite in the colour circle are used; these are red–green, red orange–blue green, orange–blue, orange yellow–blue purple, yellow–purple, green yellow–purple red. Better harmony can be obtained by mixing a little of one colour with the other. Split complementaries are those colours which are almost opposite in the colour circle, i.e. green and red orange or purple red; orange with blue purple or blue green, and so on.

Harmony of sequence or analogous harmony is when 3 or more colours or tones of colour next each other in the colour circle are used, i.e. orange-red, red orange, reddish orange orange; green, yellowish green, green yellow, greenish yellow, yellow. Contrasting harmony is when a light colour is used with a dark colour giving contrast of tone, such as pale yellow with deep blue, or, at its fullest extent, black and white.

Polychrome and triad schemes are based on the modified use of the pigmentary primary colours, i.e. a large area of a tint of one of the primaries such as cream, with areas of red at full strength and slate blue; or large areas of two modified primaries with the remaining colour full strength. Such schemes need to be well balanced in area, tone and intensity. Colours are either light or dark, bright or dull. Brightness or dullness is referred to as chroma or tonal value, bright being high and dull low. Other schemes would be the use of three or four colours kept in control with black or white. The black would give a richer but deeper effect; the white add sparkle. Warm colours are those suggestive of fire and sunlight, extending from red-purple to yellow. Cool colours are those which are suggestive of water and foliage, and extend from blue-purple to green.

Harmony can usually be secured if the colours follow the natural order of progression from light to dark, of which there are two spheres, a red and a green. The first goes from yellow the lightest through yellow, orange, orange-red, red, red-purple to purple the darkest. The second from yellow through yellow-green, green, blue-green, blue, blue-purple to purple. Thus, if yellow and blue-green are used together, the yellow should be the lighter of the two. If this natural order of colour is reversed, then the result will be discordant. When a colour is changed by the colour alongside it, this can be remedied by adding to the changed colour some of the colour causing the change.

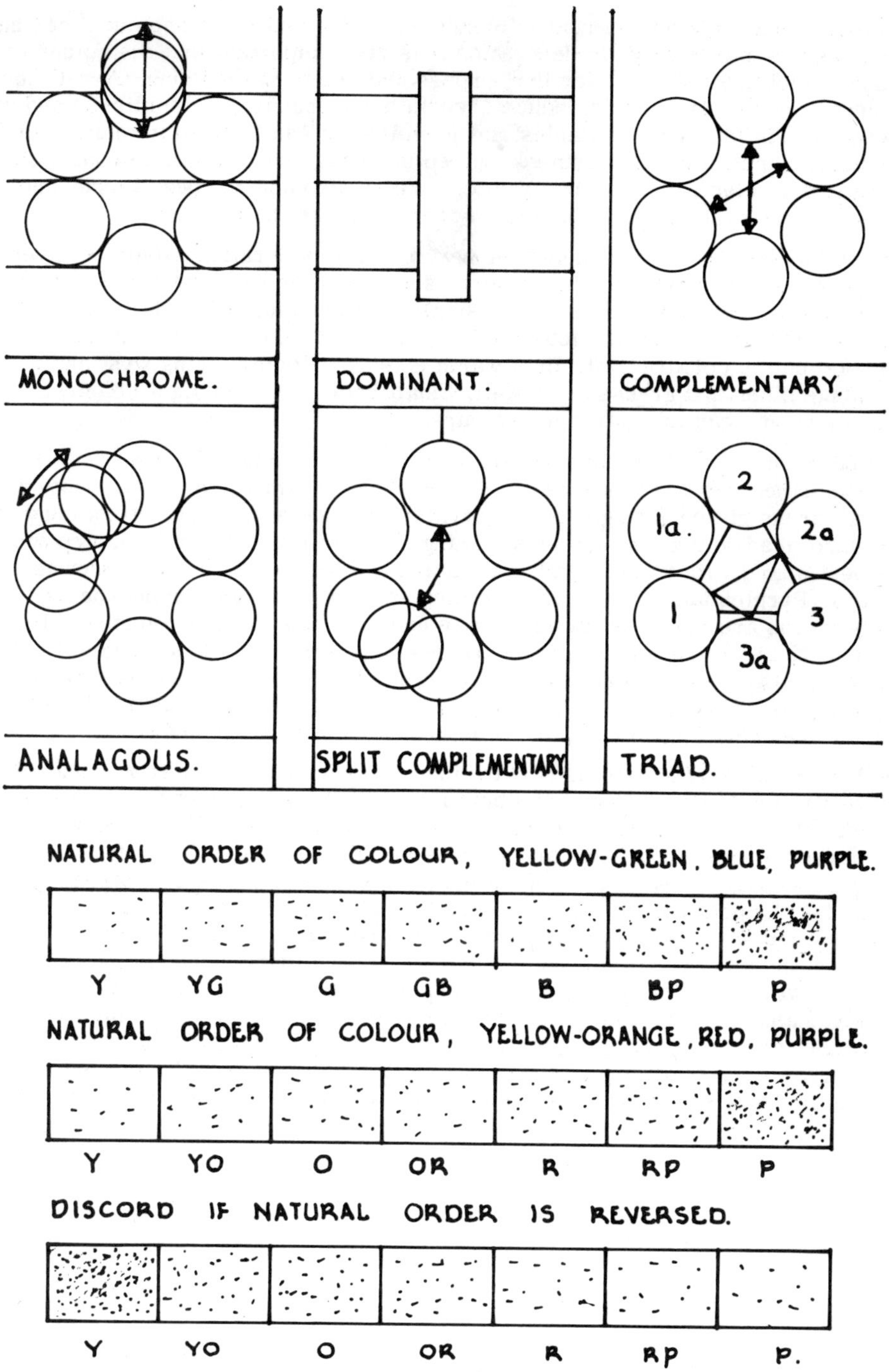
MONOCHROME.
DOMINANT.
COMPLEMENTARY.
ANALAGOUS.
SPLIT COMPLEMENTARY.
TRIAD.
1a.
2
2a
1
3
3a
NATURAL ORDER OF COLOUR, YELLOW-GREEN, BLUE, PURPLE.
Y YG G GB B BP P
NATURAL ORDER OF COLOUR, YELLOW-ORANGE, RED, PURPLE.
Y YO O OR R RP P
DISCORD IF NATURAL ORDER IS REVERSED.
Y YO O OR R RP P.

COLOUR III.

Colour IV

Colour has always had a great influence on people and since ancient times has been used symbolically. In decoration it is very important that the colours chosen should be suitable for the purpose and use of the building. A well chosen colour scheme can inspire, induce cheerfulness, encourage cleanliness and well being, and, in factories and industrial premises, it can be used to make seeing easier, promote mental alertness, and so increase efficiency and production. By careful selection it can be used to make small rooms appear larger and large rooms more cosy; to brighten rooms or make them restful.

When choosing a colour scheme for any room, consideration should be given to its size, aspect—whether it faces north, south, east or west—and the use to which it is put. A room which gets only a small amount of natural light or which faces north or east should not have large areas of cold colours—blue, green or purple—but be coloured with light warm colours. Rooms facing west and south can take blues and greens quite well. Small rooms require light colours; larger rooms can take stronger colours.

Colours which stimulate and excite, such as yellows and reds, are useful for schools, theatres and restaurants. Large areas of white and broken white with small areas of red, orange and yellow also produce stimulating effects and can be used to advantage in factories. Orange is the warmest of all colours and is best mixed with white; at full strength it can be used to liven up browns and greys. Purples and mauves are soothing and very suitable for bedrooms. Soft greens are very restful and can be used for bedrooms and lounges. Browns are restful and warming but need touches of yellow and orange or gold to cheer them up. Light greys mixed with red, orange and yellow are suitable backgrounds or foils for brighter colours. Black need not be depressing if used in small areas and can often sharpen and steady a colour scheme.

Halls, passages and staircases providing a link with all other rooms should have a warm friendly aspect and not clash with the colours of the rooms. A useful plan is to have large areas of soft warm colours with smaller areas of darker or greyed colours and accents of bright strong colour. It is useful to have important features such as the main entrance doors in a strong colour; likewise the process can be reversed, the wall being a strong colour and the doors a much lighter colour. Living rooms require colours which form a good background and are suggestive of comfort. The colours should be quiet and restful without being dull. Kitchens should be light and bright, suggestive of hygiene and cheerfulness. Public rooms like restaurants, cafes, cinemas and theatres, which are occupied for a comparatively short time can be given stronger colours than private rooms. To assist decorators the British Standards Institution has issued a series of 101 colours suitable for use in buildings.

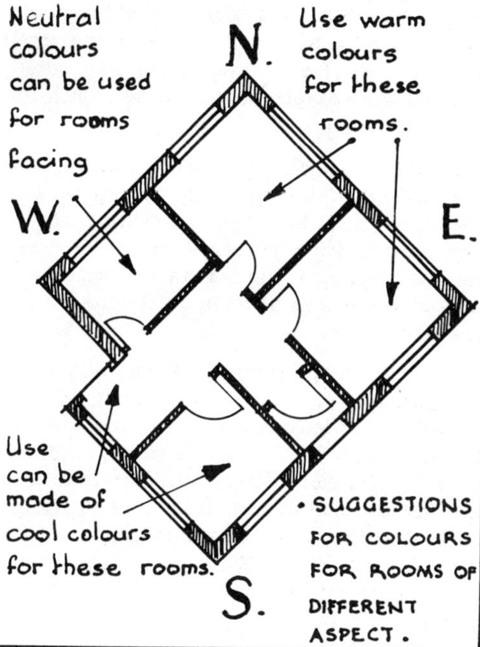

COLOUR IN INDUSTRY.
Identification of services by colour - B.S. 1710 - 1951.

SERVICE.	B.S. COLOUR. B.S. 381.C.
water.	french blue 166. or grass green 218.
steam.	aluminium.
air.	white.
drainage.	black.
electricity	light orange. 557
coal gas.	canary yellow. 309
oils.	light brown. 410.
chemicals	light grey. 631.
fire installations	signal red 537.

COLOUR IV.

Paint Spraying Equipment

The application of paints and industrial finishes by spray has been developed considerably during the last 30 years. The usual method of spraying is by petrol-driven or electrically operated compressors which deliver the necessary volume of air to the pressure pot or spray gun; and, although it is usual to have static air compressors in factories, with building and decorating work portable equipment is necessary. Electrically driven compressors are most useful, but where the supply is unsuitable or non-existent, as at a new building site, a petrol driven compressor will be needed.

Spray compressors are of two types: (a) 'air volume', which use a large quantity of air at low presure, often as low as 30-48 kN/m^2, (b) high pressure, which use a smaller volume of air at 400-550 kN/m^2; and then the air system governs the rest of the equipment. Low pressure systems use either 0·25 or 0·5 litre gravity feed cups fitted to the top of the spray gun or, for larger quantities of material, a pressure pot. High pressure systems use suction feed cups fixed to the underside of the spray gun or pressure pots.

The simplest type of outfit consists of a gravity cup fixed to a spray gun with a single air line delivering air from the compressor to the gun, which, passing through the gun, sucks the paint out through the nipple and so on to the surface. A more elaborate outfit has a pressure pot which holds 10-200 litres of material which is pushed up a fluid hose or line to the gun, with an air line to complete the outfit. High pressure equipment for small jobs consists of a suction cup fixed to the bottom of the gun and an air line from gun to compressor. For large volumes of work, a pressure pot with fluid hose and air hose are essential. An air transformer on the machine regulates the pressure of the air and acts as an air filter.

Spray guns are of two main types: (a) the bleeder type, which is used with a compressor without pressure control. They have an intentional air leak from some part of the gun, to prevent air pressure being built up in the hose, and the trigger of the gun only controls the flow of the material; (b) the non-bleeder type, which is fitted with an air valve which shuts off the gun when the trigger is released. The trigger thus controls both air and material. The nozzle of the gun needs to be varied for applying various types of material. Light weight materials and very fine work need an aperture of 0. 8-1. 2 mm. For general work, the standard size is 1. 8 mm, and for distempers and similar mixtures 2. 2 mm. Thus, the best type of spray guns have removable heads which, by changing the head, enables all types of material to be applied with one gun.

To convert kN/m^2 into lbs per sq. in., divide by 7.
This gives an approximate conversion.

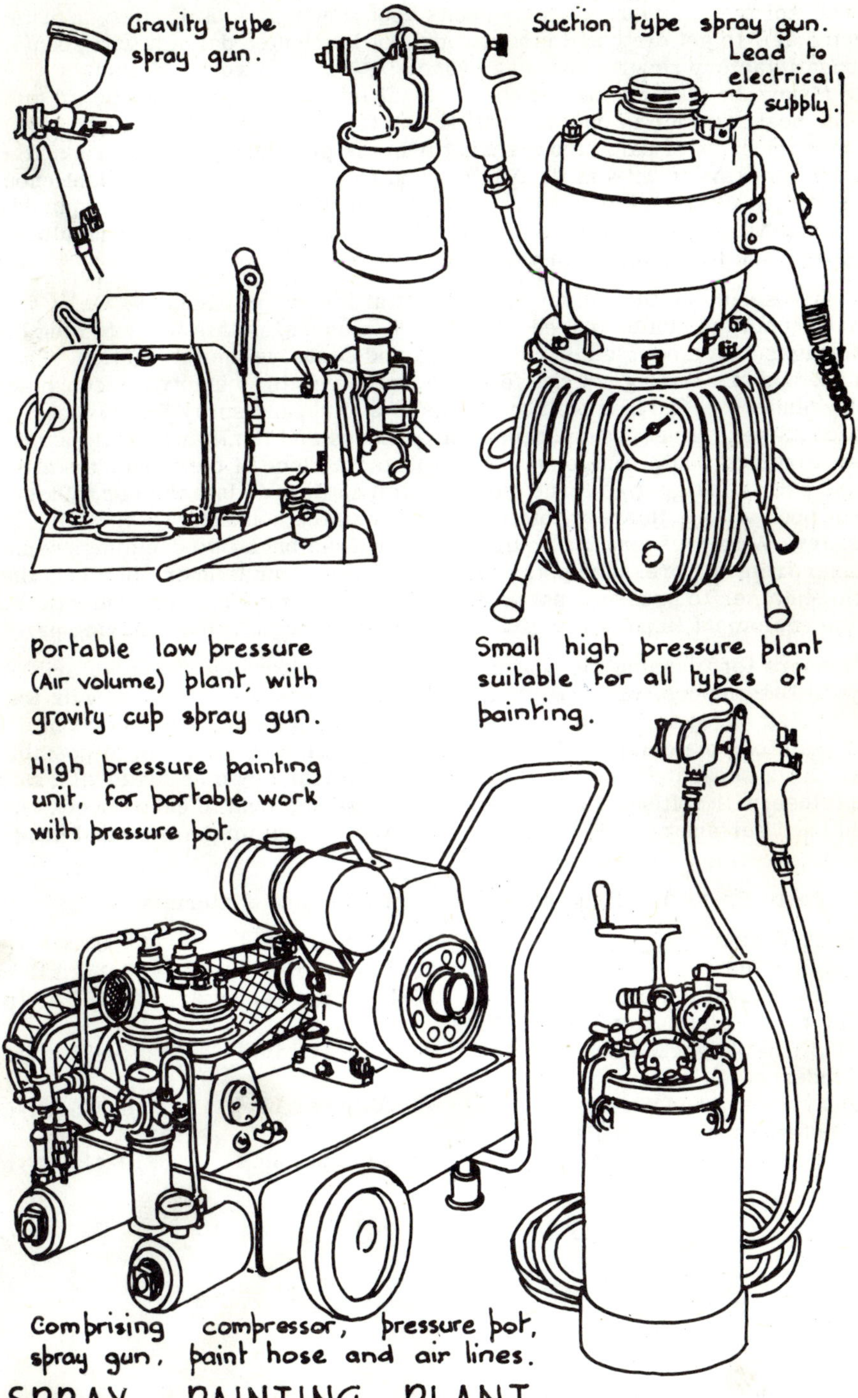

SPRAY PAINTING PLANT.

Care of Spray Equipment

How well spray equipment works depends upon how it is looked after. If it is neglected, left to get dirty and clogged, and not lubricated daily, it will soon give unsatisfactory results and may eventually cease to work. Materials should always be strained before they are put into cup or pressure pot. For cups, the best strainer is one that will fit the top of the cup. The practice of opening a fresh tin of paint and standing it in the pressure pot is all right if there are no skins or nibs in it. All nuts and connections should be tight enough to do their job effectively but not so tight that unnecessary effort is required to undo them. All parts with knurled edges should be 'finger tight' and no attempt should be made to tighten them with pliers.

Spray guns should be thoroughly cleaned out at the end of each day. On no account should the gun be placed in a dish of thinners, as this tends to clog the air passages, dries out glands and packing, and destroys the lubrication. The usual method of cleaning gravity feed cup equipment is to empty the cup, place a little thinners in it, rinse well by means of a fitch, and spray the thinners into a container. If necessary, repeat the process and finally wipe with a rag. For suction feed equipment, empty the cup, hold a piece of cloth over the air cap, and pull the trigger to force any paint left in the gun into the cup. Wipe the cup, pour a little thinners into it and spray into a container. Repeat if necessary. With pressure fed equipment, blow fluid back into container, remove container from the pressure pot, empty and clean. Pour thinners into container, return container to pressure pot and blow through gun. Wipe tank and gun. Before the equipment is put away give a little lubricating oil to all moving parts.

The thinners for cleaning the equipment will vary according to the materials sprayed. Distempers, water paints and cement paints can be cleaned with warm water. Spirit varnishes need methylated spirits; oil paints, paraffin or white spirit; cellulose materials, cellulose materials, cellulose thinners. If possible, always use a separate hose for each type of material, as some materials, such as cellulose, will soften oil paints left in the hose and tend to cause specks on the finished surface or clog the spray gun. Always coil up the hose and hang them up.

Approximate Pressures Required for Spraying Various Materials

Spirit Stains	12-15 lb per in.^{2}	83-103 kN/m^2
Spirit Paints	20-25 lb per in.^{2}	137-172 kN/m^2
Light-Weight Materials (Flat Oil Paints; Aluminium Paints)	20-25 lb per in.^{2}	137-172 kN/m^2
Medium-Weight Materials (Undercoats)	30-40 lb per in.^{2}	207-276 kN/m^2
Heavy-Weight Materials (Distempers: Cement Paints)	45-50 lb per in.^{2}	310-345 kN/m^2
Cellulose Materials	45-60 lb per in.^{2}	310-414 kN/m^2

In the table above pressures are quoted first in Imperial units, since pressure gauges are still generally calibrated in lbs per sq. in. SI units are shown for comparison.

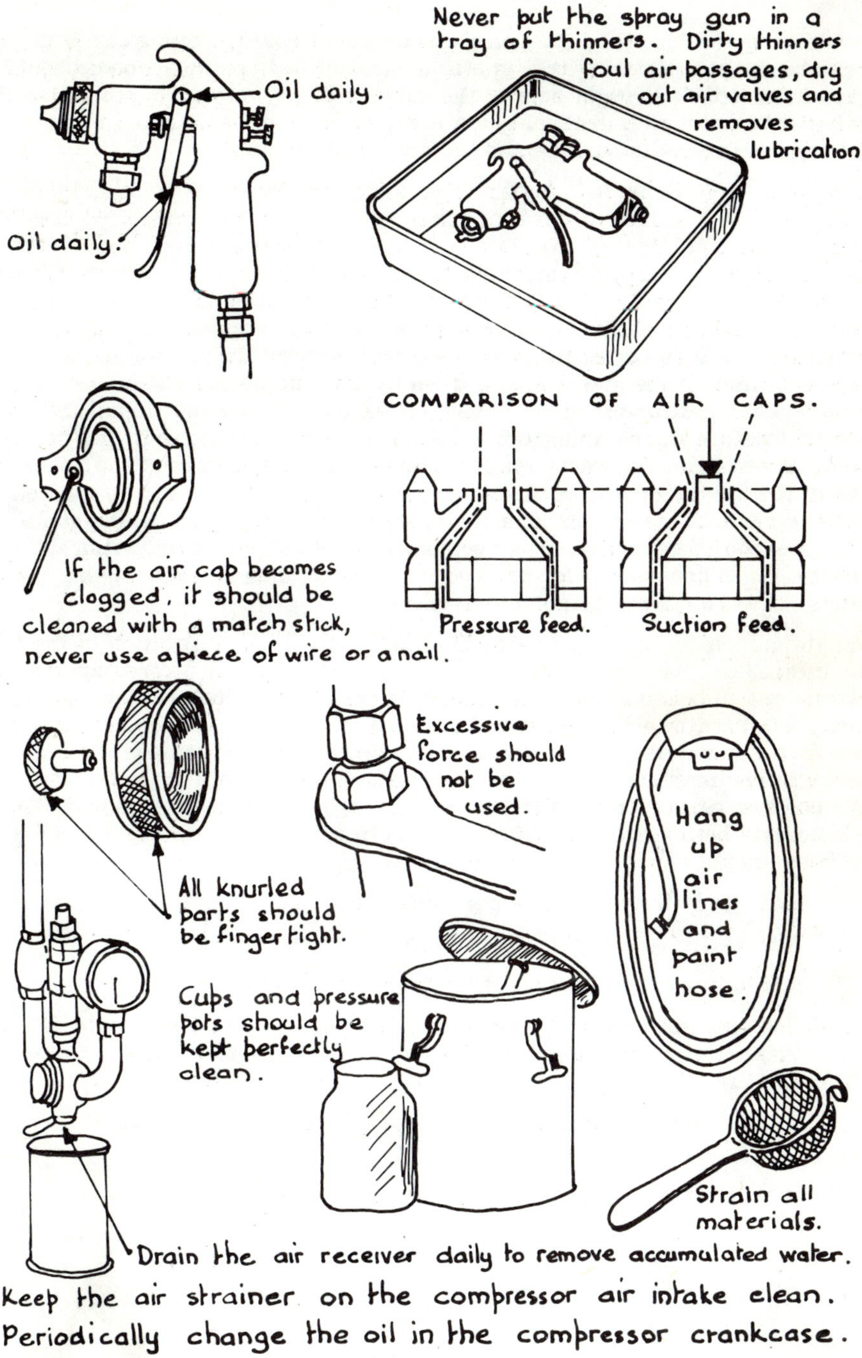

SIMPLE CARE OF SPRAY EQUIPMENT.

Spraying Techniques

When spraying, the nozzle of the gun should be held 150-200 mm away from the surface and the gun should travel in a line parallel to it; on no account should the gun be 'arced' or brought nearer the surface at any one point, otherwise the coating will be patchy. A general guide for distance can be obtained by stretching the hand between the surface and the nozzle of the gun.

When doing a large flank, it is better to go all round the edges first and then treat the area as a series of vertical strips about 1 m wide, carefully overlapping each strip to ensure adequate cover. Horizontal surfaces should be done edges first and then, commencing at the near edge, sprayed in a continuous movement until the whole area is covered. When spraying an external angle, the gun should be aimed directly at the edge, so as to catch both sides, but for an internal angle it is better to finish each face separately, as this gives a more even finish. If the gun is aimed directly into the corner, the material piles up on each side and tends to cause patchiness. It is usual to pull the trigger just before the commencement of the stroke and release it just before the end of the stroke. This economises on material, cuts down mist and fumes, and is less tiring for the operator. Round and curved surfaces should be treated as a series of vertical strips and no attempt should be made to move the gun to follow the shape of the curve. For open work, such as grilles or railings, the aim should be to cover as much area as possible with each stroke, using three or four strokes to ensure complete coverage.

By simple movement of the diffusing ring or air baffle, the direction of the jet can be altered to give a vertical pattern, a horizontal one or a circular one. If the nozzle gets blocked and the jet pattern becomes distorted, remove the nozzle and clean it with thinners and a sharpened match stick; on no account use a piece of wire or nail for this purpose; otherwise it might damage the hole. For very intricate surfaces, such as decorative plaster work or radiators, spray guns are very useful tools; and they can apply metallic paints and cellulose materials better then either brush or roller. To obtain the best results, the painter should:

(1) Make sure the surface is perfectly clean and smooth.

(2) Always spray at right angles to the surface.

(3) Maintain an even speed, not too hurried or too slow.

(4) Use only sufficient air pressure to properly atomize the material. Too little or too much air pressure causes roughness and an 'orange peel' finish.

(5) Overlay all strokes at least ⅓ to produce an even coating.

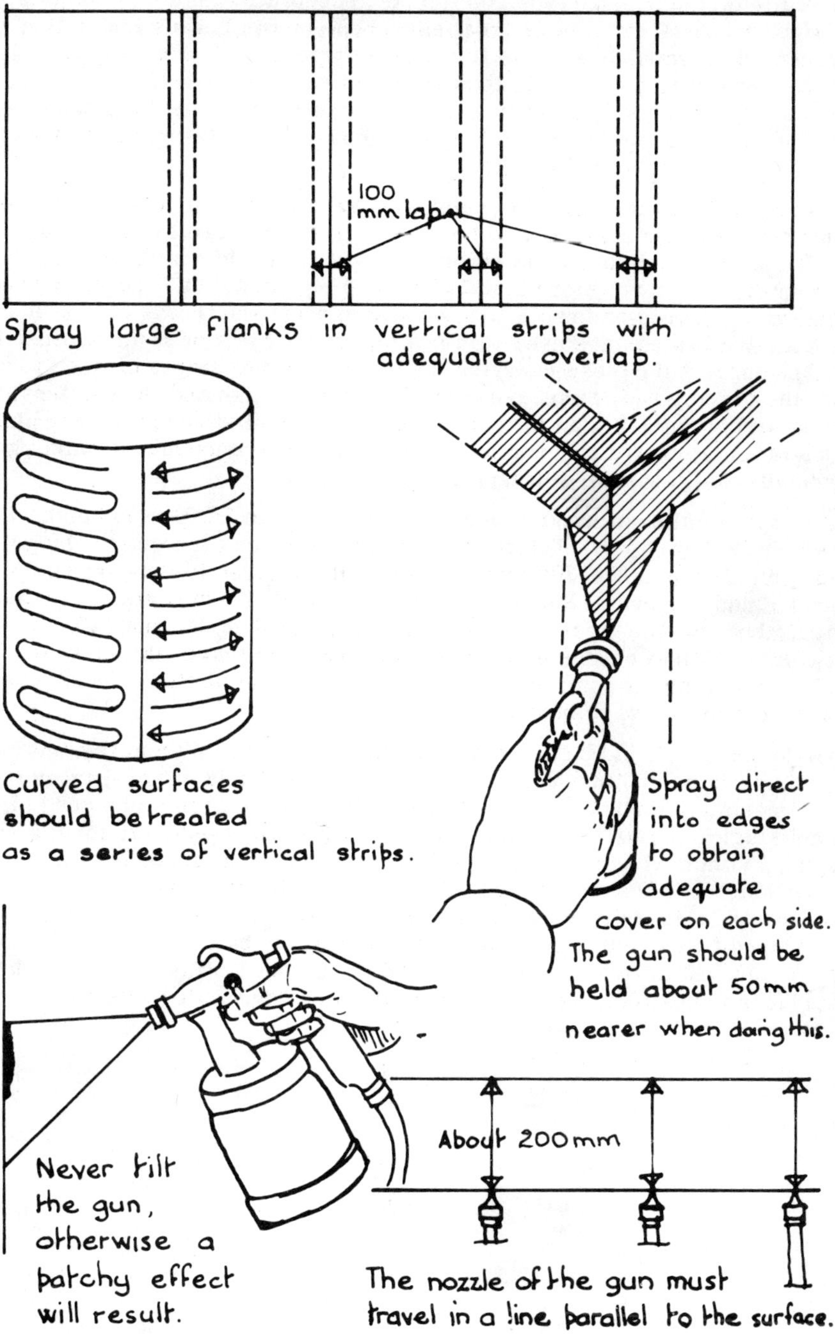

SPRAYING TECHNIQUES.

Scaffolding

STEPS, TRESTLES, BOARDS, STAGING

The scaffolding normally required for the redecoration of the inside of a small house is two pairs of steps or trestles and one or two boards for ceiling work, with perhaps a small ladder 10-12 rungs for a long wall on a staircase. Larger rooms, where it is not suitable to keep on shifting a step scaffold, will require a running scaffold made up of two runs of ledgers supported by a suitable number of trestles so that the scaffold boards or lightweight staging units can be moved along as the work proceeds.

Steps are of metal or wood, the metal ones being lighter in weight and less liable to damage. Wooden steps should be made of straight grained timber with treads of sufficient strength and wide enough to take a scaffold board, with the distance between the treads not more than 175 mm. They should be securely hinged at the back, with cords or braces long enough to allow the treads to rest in a level position. When steps are used with scaffold boards in a built up scaffold, it is usual to support the tread carrying the board with two pieces of wood wedged from the lower tread. Steps are made 5-16 treads high and for painters' work they should be lighter than for general building. Tall steps or moving platforms with a platform, guard rail and tool tray at the top are useful for painting isolated units such as lamp standards, etc.

Trestles are either of wood or metal and range from 1·8-6 m in height, with bars 450-600 mm apart. The bars are either level on each side or staggered to give an alternate 225-300 mm rise; with wooden trestles they should be mortised and tenoned to the stiles. Trestles should be wide enough to take two scaffold boards side by side, giving a working platform at least 425 mm wide. The platform must never be higher than 4·5 m from the ground on which the trestles stand, and if scaffold boards are used for the platform there must be a double thickness of boards.

Scaffold boards should be 200 mm wide unless they are 50 mm thick, in which case they can be 150 mm wide, but no less. They should not project more than four times their thickness unless adequately fastened at the other end. Lightweight staging consisting of strips of timber braced together to form a rigid staging 450 mm wide can be obtained in lengths ranging from 2-8 m. A simple device which utilizes scaffold boards on edge as ledgers are split-heads, which consist of a telescopic column fixed on three legs. At the top of the column is a U-shaped fitting, into which the edge of a scaffold board can be fixed. With the columns adjusted to a suitable height (less than 3 m) pairs of split-heads are placed in position, the boards being fixed on edge with cross boards to give a working platform.

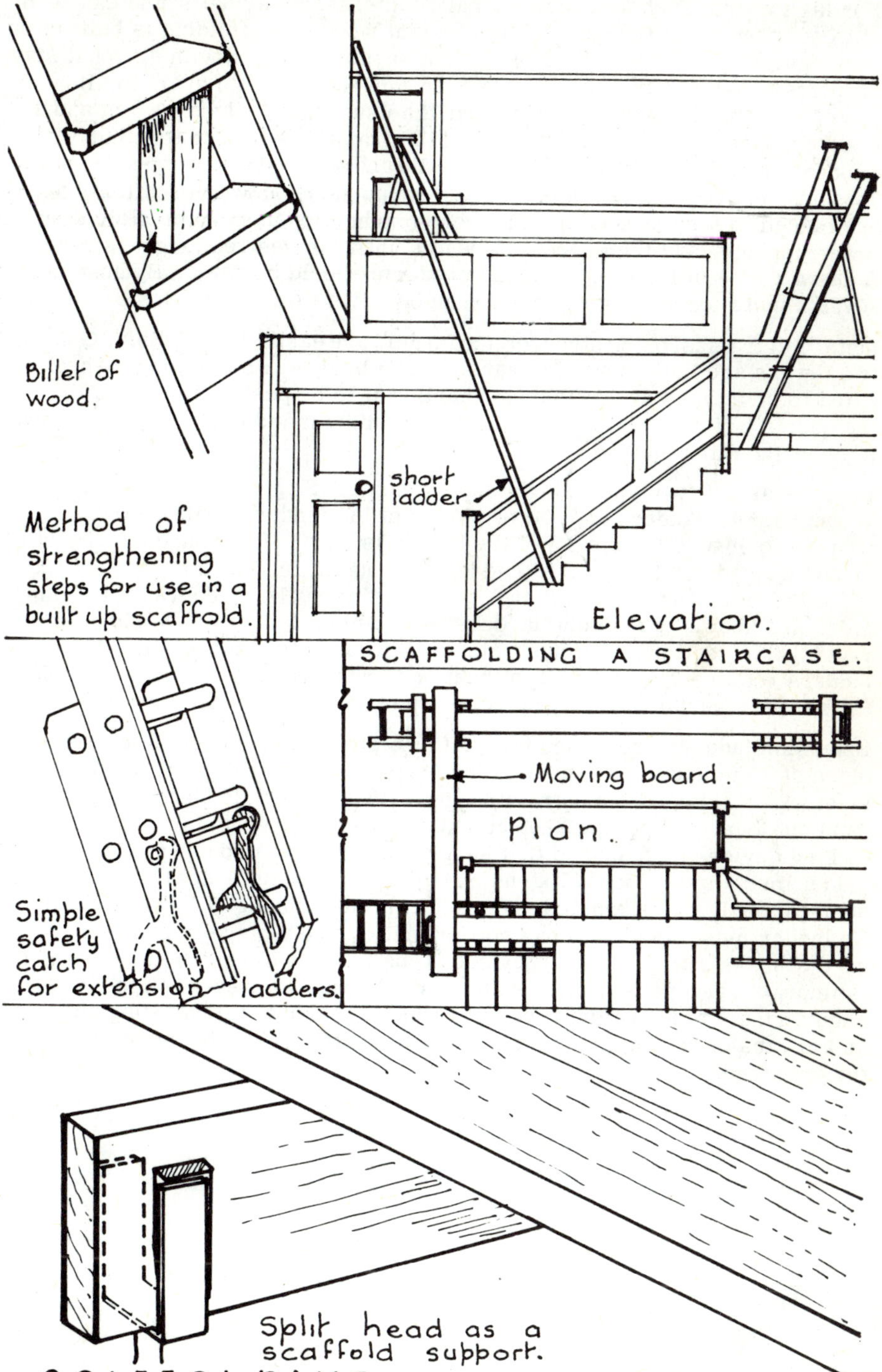

SCAFFOLDING.

Scaffolding

WOOD AND METAL STAGING, LADDERS

For high rooms such as churches, ballrooms, assembly halls, use can be made of wooden or metal framed staging. Patent wooden scaffolding is built up in rectangular frames by means of wooden supports braced with diagonal stays. Supports consist of two timbers 75 × 25 mm fixed 40 mm apart to allow the end of a scaffold board to be inserted, the usual lengths being 1·8 m and 1 m. Four supports forming the corners of the frame are braced together with wooden struts 3·5 m long, fixed diagonally and secured with wing-nuts and bolts. Two boards are inserted in the slotted supports, upon which scaffold boards are placed to form a staging. Further upright supports can be slotted into the lower uprights, and the process repeated until the desired height is reached. A guard rail 1 m from the working platform should be fitted. By assembling several units together, large areas can be covered.

Metal unit frames, being stronger and larger, are easier to assemble. Generally, two towers are built, boarded over and then bridged with staging. Wheels can be fitted to both types of scaffolding, enabling it to be pushed along, and this enables the quantity of scaffolding to be cut to a minimum and part of the floor space to be utilized if required.

Ladders are of two types, single and extension. Single wooden ladders are known as pole-ladders and the two sides of the ladder should be from the same fir pole to ensure the spring of the ladder is equal. Rungs are usually of oak or ash spaced 250 mm apart. Ladders are reinforced with steel rods fixed under the top and bottom rungs and at suitable intervals. Ladders with rectangular sides are known as standing ladders. Aluminium ladders have sides of hollow rectangular section and rungs which are grooved to prevent slip. These ladders require spikes fitted to them on frosty ground or rubber suction feet for slippery surfaces. They do not warp or rot like wooden ladders.

Extension ladders are either two or three piece;those about 10 m are 2-piece, those of 15 m and over 3-piece. The backs of the sides are rebated and stout steel wires inserted for extra strength. Most 2-piece ladders are pushed up by hand but 3-piece and long 2-piece extensions are operated by ropes with special locking devices so balanced that should the rope break during extension, the ladder immediately locks. Extension ladders are useful, since the sections can be used singly or assembled, thus reducing the number of ladders required. No ladder must stand on loose soil or bricks, and a piece of board should be used to make a firm base. They should be fixed at top and bottom, but if this is impractical a person must stand at the bottom to prevent slipping. They should rise 1 m above the rung upon which the painter is standing and, if used as a means of getting from one staging to another, rise the same distance above the top stage.

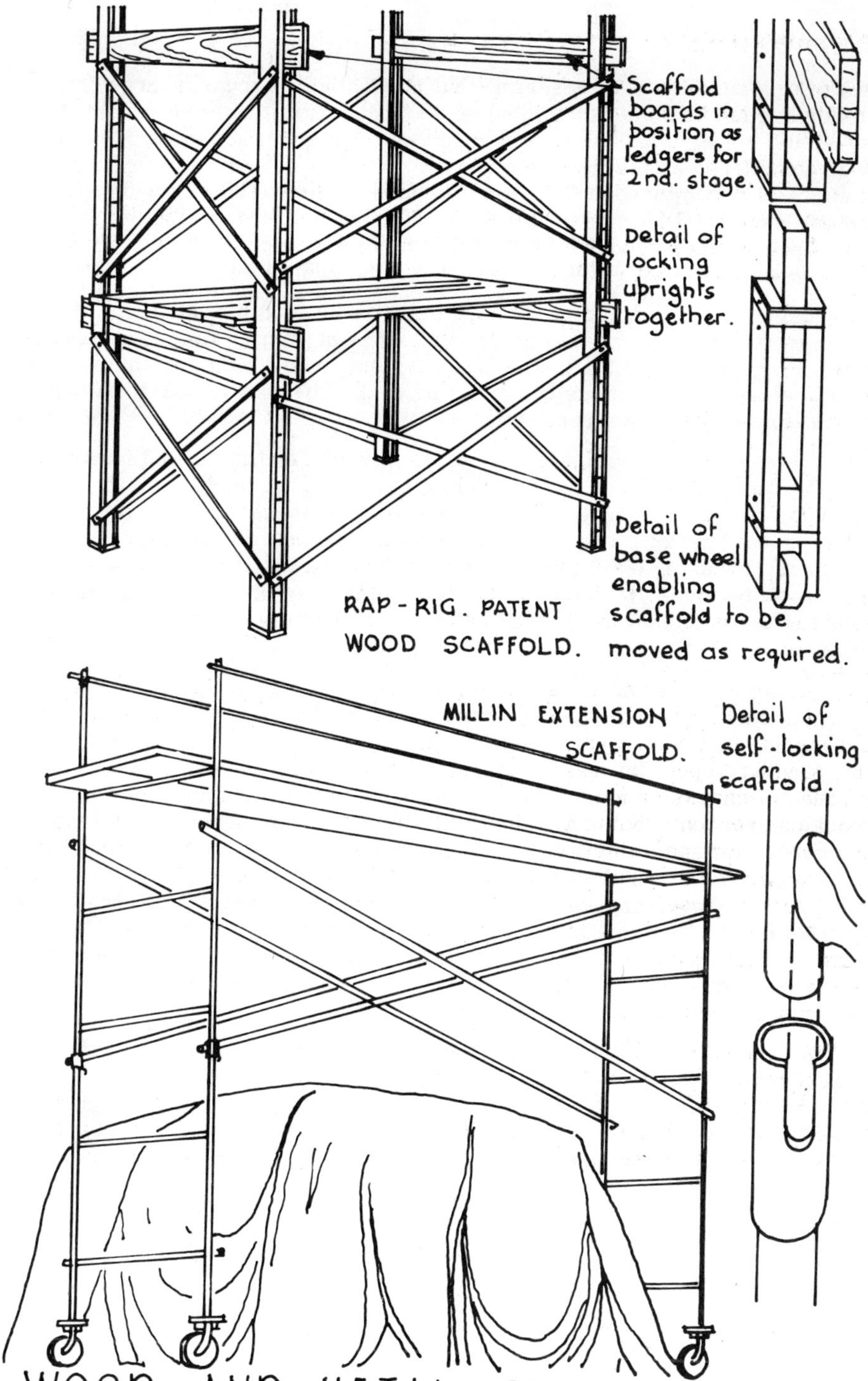

WOOD AND METAL STAGING.

DUCK-BOARDS AND SAFETY NETS

When working on roofs it is essential that adequate duck-boards, crawling boards or hen ladders both in number and construction should be provided. These should not be make-shift ones made up from odd pieces of timber but must be considered as pieces of equipment just as necessary as steps or trestles. It is recommended that they should be at least 400 mm wide with a maximum length of 3·75 m; constructed of 50 × 25 mm timbers bolted at one end to a head or anchor board 400 × 130 × 50 mm reinforced with a chamfered member to keep it rigid for fixing over the roof ridge.

The treads or battens should be 30 × 25 mm fixed at 400 mm centres. If necessary extra boards can be coupled by means of 10 mm diameter steel pins inserted near the ends of the board in the two innermost timbers and connected with steel straps. Metal skids can be fitted to the boards to make it easier when positioning them up a roof slope.

Metal ladders are also available, the staves of which end in curved hooks which engage the roof ridge. They have wheels for pushing the ladders up the roof slope. Where it is not possible to secure sufficient anchorage with the anchor board, or in the case of metal ladders the hooks, the boards should be secured by means of ropes. At least two sets of crawling boards are essential to enable shifts to be made. It is unsafe to step on the area of the roofing where it is bolted to the purlins during shifting, as some roofing sheets shatter without warning.

When working under roofs in factories involving risks from dangerous machinery or fumes, or on jobs such as railway termini, hangers, and under bridges or at heights of 9 m or over, safety sheets are hung below the scaffold. There is no standard specification for these sheets but they are usually 9 × 6 m and made of canvas or nylon. The best types are washable and fire-resistant for obvious reasons. Some have diagonal strips for reinforcement and the tying ropes sewn onto the sheet at 1 m intervals. Others are reinforced with nine 150 mm wide strengthening bands with 20 mm-diameter steel rings inserted into the canvas at 1·8 m centres along the length and 1·5 m centres along the width. The ropes for fixing are wire heart hemp and the method of jointing to form the loop tuck-spliced. Open stitching is to be preferred as the strength of the sheeting is retained, whereas with close stitching a sudden impact may cause the sheet to tear.

The ropes should be fixed round the poles in such a way that they do not trip the workers and the sheets should be sufficiently slack to enable them to be easily distinguished from boards. In situations where heat is liable to twist scaffold boards and render them dangerous it is better to use fir-poles for ledgers rather than tubes, so that the scaffold boards can be secured by nailing.

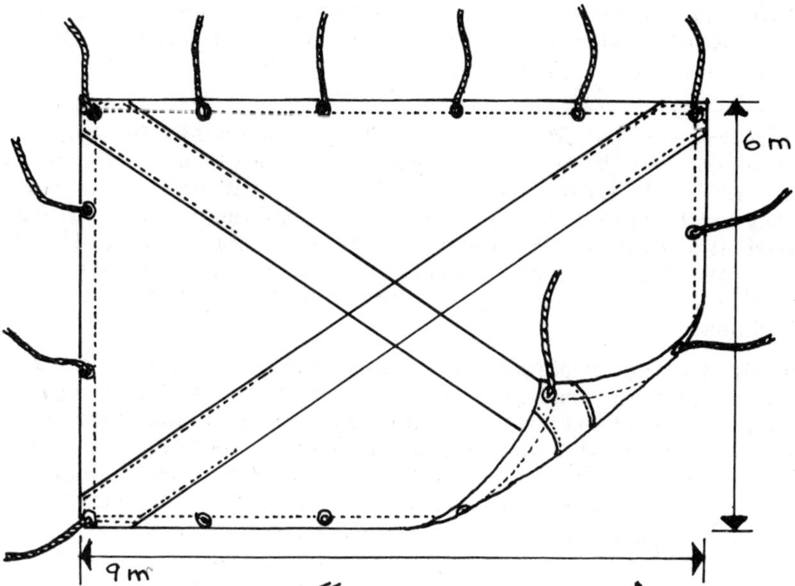

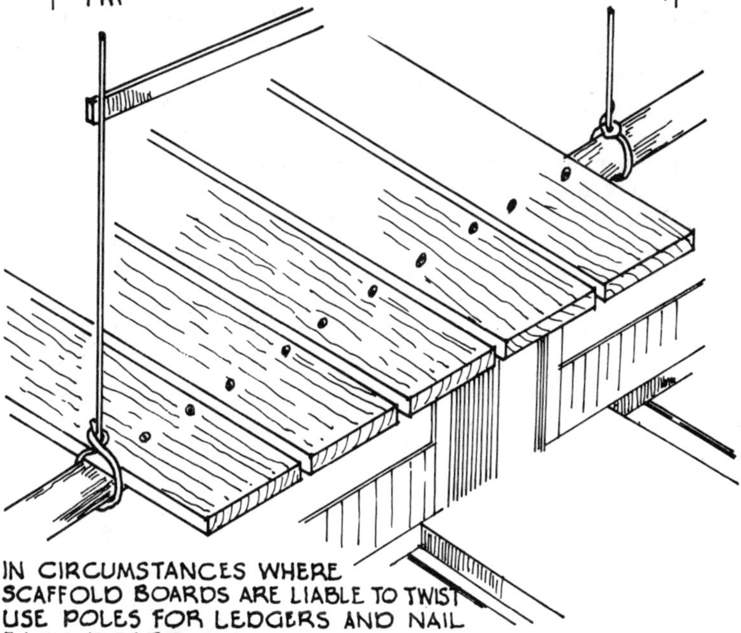

IN CIRCUMSTANCES WHERE SCAFFOLD BOARDS ARE LIABLE TO TWIST USE POLES FOR LEDGERS AND NAIL EACH BOARD.

SAFETY SHEET AND SLUNG SCAFFOLD.

LADDER BRACKETS, CRADLES AND BOSUN'S CHAIRS

Ladder brackets are used for temporary scaffolds for outside work. They consist of a wrought iron bracket with an adjustable bar to make a level support and are clipped on to two rungs of a ladder. They can be obtained for use above or below the ladder. The usual form is to have two ladders with brackets at a suitable distance apart with scaffold boards forming the platform. Cripples, which fix below the ladder, are of similar form but with a wooden platform from which the painter works.

On buildings where it is impracticable to use a built-up scaffold or ladders, as, for instance, in the well-hole of a block of flats or in a busy thoroughfare, cradles are essential. There are two types: (1) those which can only be raised or lowered vertically; (2) those which travel along a rail fixed to out-riggers so that they can be moved along the length of the building and vertically as well. Cradles are either wooden or fibre-glass with metal stays, usually 1·8 m long and 0·7 m wide. Being a suspended platform, it is necessary to have a guard rail fixed at a height of 1 m and a toe board to guard against tools being kicked or knocked down.

Fixed cradles are hung from pulleys fastened with wire ropes or 'bonds' on the end of poles cantilevered on the roof with two or three 25 kg weights; and the cradles are raised and lowered with manilla ropes 70 mm in circumference fixed to the lower pulley, passing up to the top pulleys, down to the bottom pulleys, up to the top again and finally down to the cradle, where they are fixed by making two hitches in each rope around the cradle stays. By untying the bottom hitches and pulling on the ropes, the cradle can be raised and retied, or, by letting the ropes slip slowly through the bottom blocks, the cradle can be lowered. Travelling cradles which work along a rail or track are manipulated in the same manner, the horizontal moves being made by means of a third rope fixed to the rail.

When only isolated units such as a rainwater stack or chimney have to be painted, use can be made of a bosun's chair. This is a seat 0·5-0·7 m long and 225 mm wide, suspended in the same manner as a cradle but from a single point. A back rest, 250 mm above the seat and a central forward member enabling the person using the chair to put one leg on either side of the member, must be provided. The point of suspension should be at least 0·5 m above the seat and be so connected to it that the chair is stable under all conditions.

STORAGE OF SCAFFOLDING

All wooden scaffolding and equipment should be stored under cover, short ladders, steps and trestles standing upright, scaffold boards and long ladders laid horizontally well above ground level.

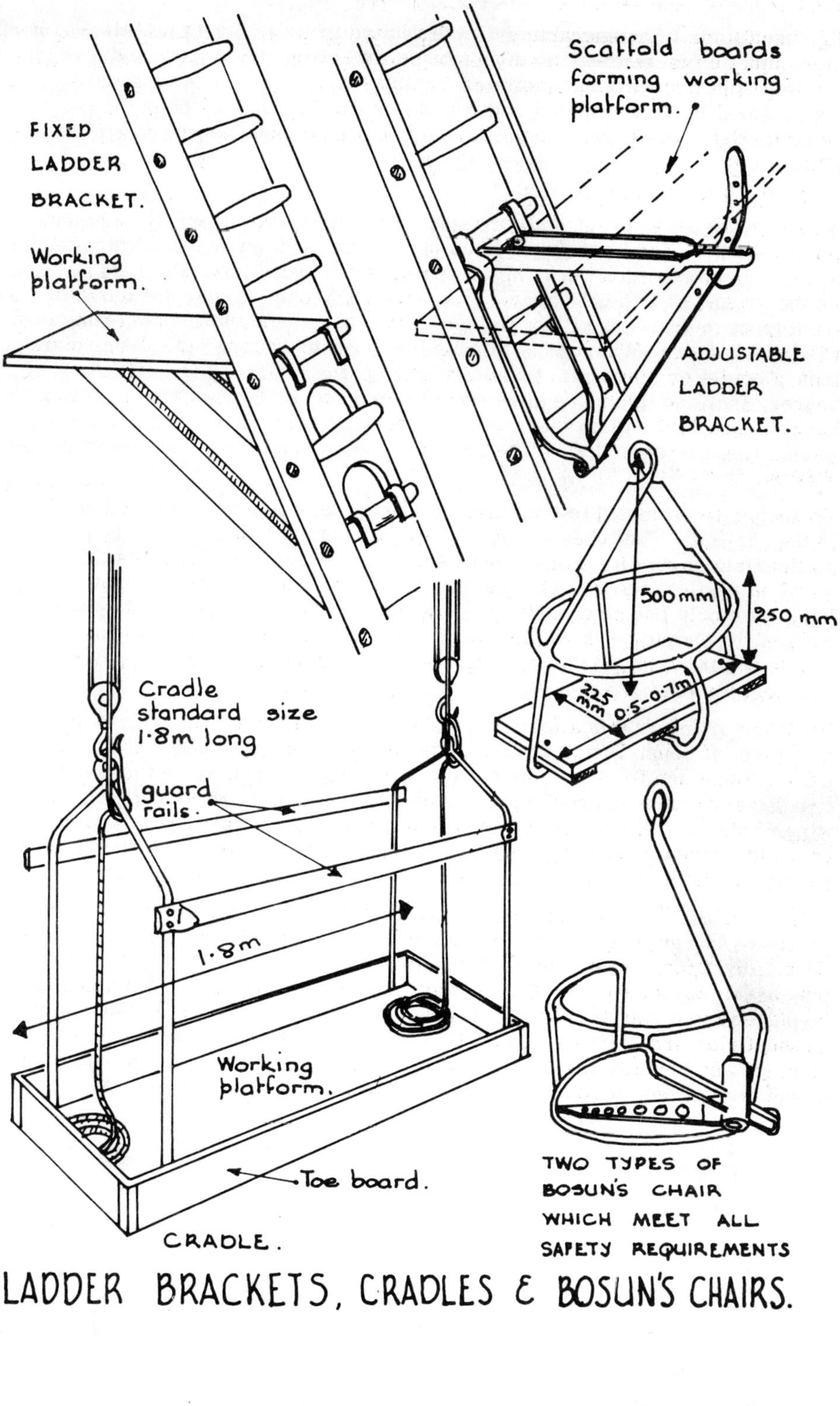

LADDER BRACKETS, CRADLES & BOSUN'S CHAIRS.

TYPES OF BUILT-UP SCAFFOLD

For scaffolds of considerable size which have to be in position for some weeks, they must be inspected by a competent person every seven days and a report kept of each inspection. Sometimes fir poles are used, and the standards or uprights stood in barrels filled with rubble or sand to anchor them, the poles being lashed together with manilla ropes or wire bonds kept tight with wooden wedges.

Poles have largely been replaced by steel or alloy tubes, which are less cumbersome, easier to erect, and stronger. The tubes have generally a 38 mm inner diameter and are spoken of as 50 mm tubes. Upright tubes which rest on the ground and are load bearing are called standards; those which do not rest on the ground and do not carry a load are puncheons. Horizontal tubes which go from front to back are either putlogs or transoms, and those which run lengthwise are ledgers. Where it is impossible to secure fixing for intermediate putlogs, a bridle or horizontal tube is fixed to putlogs already in position as a bearer. Inclined tubes are either braces or rakers. Braces give stability and keep the scaffold rigid; rakers bear on the ground or on to any rigid structure. Reveal ties are short tubes fixed between two surfaces, such as the sides of a window.

To secure the scaffold to the building, a short tube is fixed from the reveal tie to the scaffold. The tubes are fixed together with couplings or clips, each specialist firm having its own pattern. On loose ground, the standards are stood on scaffold boards, which are termed sole plates. Single or putlog scaffolds have only one row of standards and are secured to the building by the putlogs, depending on the building for stability and support. Independent or double scaffolds have two rows of standards spaced 1m apart to provide a platform for four 225 mm boards.

Birdcage or internal scaffolds have standards spaced at 3 m centres in each direction. If construction work is involved, the centres are reduced to 2 m or 2·5 m. Slung scaffolds are for ceiling work and are hung from the constructional members of the roof, or poles or tubes are laid across rafters with wire bonds. If there is an ornamental plaster ceiling underneath through which the wires must pass, provision of special holes, usually metal lined and called trumpets, is made in the ceiling so that it does not suffer damage.

Cantilever or trussed scaffolds are supported by rakers and tied to the building by means of reveal ties or tubes inside the building. Where suitable ledges occur for support this form of scaffold is useful over shops and business premises. For factory roofs which are not too high, lightweight staging can be placed from roof truss to roof truss and moved as required. Where internal scaffolds are of a fair height, 12 m and over, as at railway stations, gantries and bridges, provision should be made for the hanging of safety sheets or nets to prevent accidents.

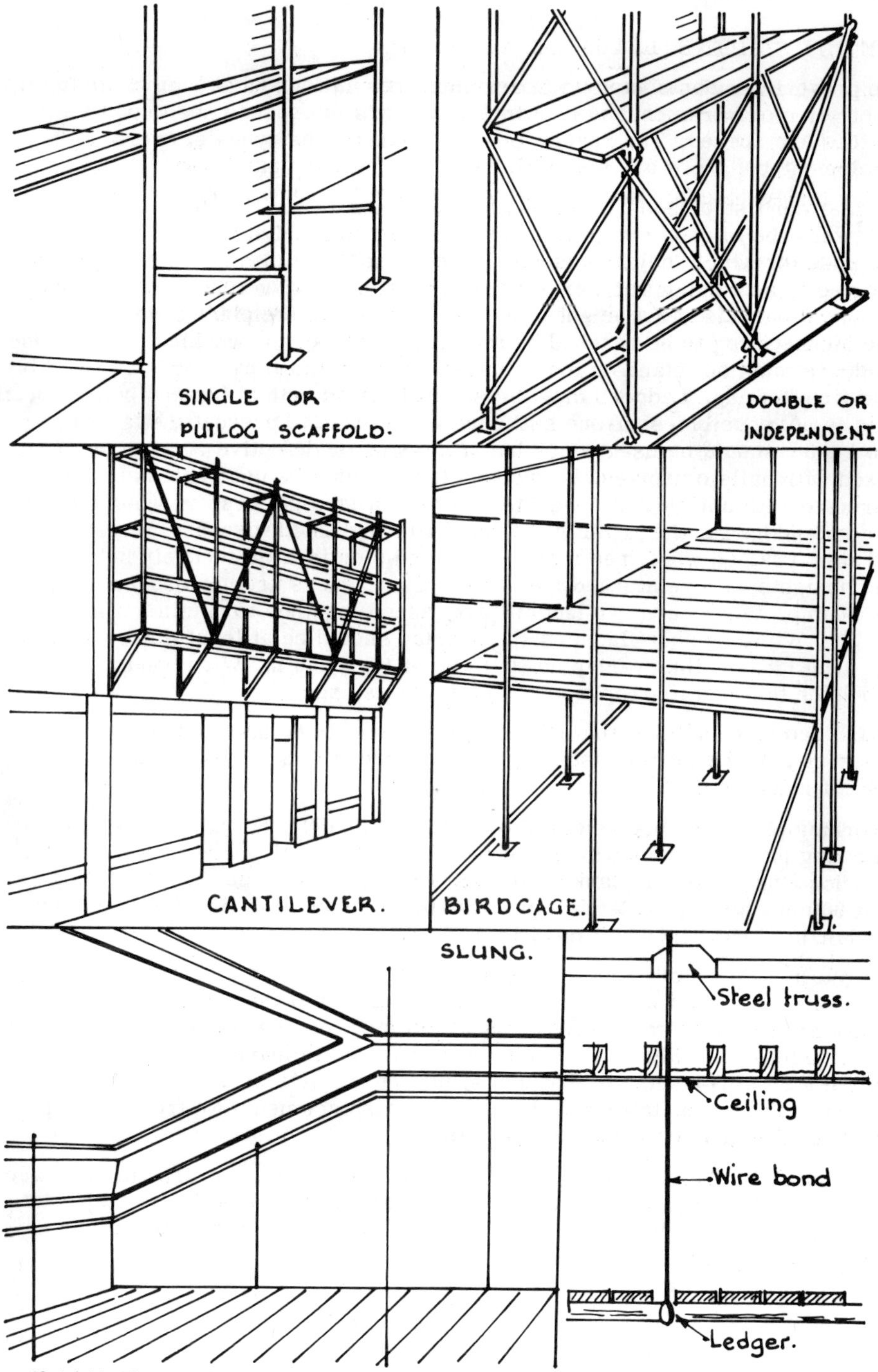

BUILT UP SCAFFOLDS.

Safety, Health and Welfare

REGULATIONS 1946 – I

To prevent accidents various Government regulations have been made for the protection of workmen, and it is in everybody's interest to see that these regulations are obeyed. The regulations cover all trades, types of operation and equipment, but there are some which chiefly concern painters.

Ladders and step-ladders are dealt with in Regulation No. 29, which states that they must be of good construction, sound material and strong enough for the purpose for which they are used. Ladders used as a means of getting from staging to staging, or for working from, must be long enough to rise, or adequate handhold provided, to a height of at least 1 m above the place of landing or the highest rung to be reached by the feet of any person working on the ladder. Ladders must not stand on loose bricks or other loose packing but have a level and firm footing. Ladders must be securely fixed both at top and bottom or, if this is not possible, someone stationed at the base to prevent the ladder slipping. No ladder should be used which has a missing or defective rung or any rung fixed with nails or spikes. No wooden ladder must be used unless it is constructed with uprights of adequate strength of straight grained wood free from defect and having the grain of the wood running lengthwise; and rungs of straight grained wood free from defect, and mortised or securely notched into the uprights with reinforced metal ties, if the tenons are not wedged. This is most important where two ladders are lashed together to obtain sufficient height. With two short ladders, the overlap should be at least three rungs, but for long ladders the overlap needs to be at least 1·5 m. Wire bonds, sash cords or thin manilla ropes are useful for lashings.

It is recommended that the foot of a ladder should be placed one quarter of the distance that the ladder rises; e.g. rise of ladder 8 m, distance from wall to foot of ladder 2 m.

Working platforms are referred to in Regulation No. 22, which states that every working platform from which a person is liable to fall more than 2 m shall be closely boarded or planked, at least 635 mm wide if used for a footing only, but 860 mm wide if materials are also to be stood on it. Where supporting a higher platform, the width must be 1·07 m.

Regulation No. 19 deals with trestle scaffolds. No trestle scaffolds shall be used if constructed with more than three tiers, or if the working platform is more than 4·6 m high. No trestle scaffold must be erected on a scaffold platform unless the width of the platform is such as to leave sufficient clear space for the transport of materials, and the trestles or uprights are firmly attached and adequately braced to prevent displacement. No trestle scaffold shall be erected on a suspended scaffold.

Regulation No. 25 states that the space between a work platform and the face of a building should never exceed 300 mm; and a clear passage way at least 430 mm wide must be left between one side of any working platform and any fixture or deposited material.

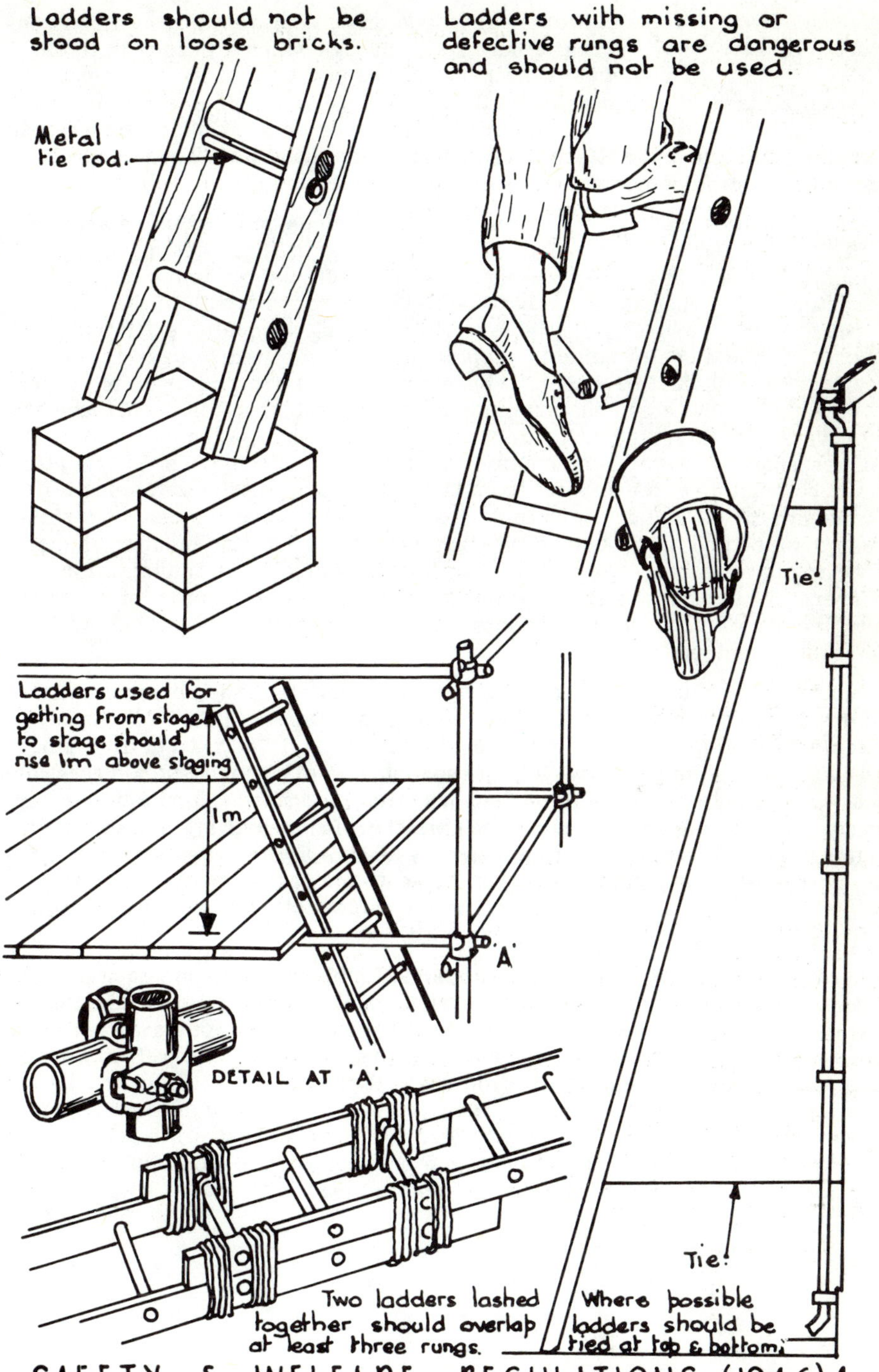

SAFETY & WELFARE REGULATIONS (1946) 1.

Safety, Health and Welfare

REGULATIONS 1946 – II

Regulation No. 31 refers to work being done on or near roofs or ceilings constructed or covered with fragile materials through which a person is liable to fall 3 m or more; it requires prominent notices stating that the coverings are fragile, fixed to the approaches to the roof. Glass roofs are not covered by this regulation. Where workmen have to pass over or work above such fragile materials, suitable and sufficient duck-boards or crawling boards securely supported must be provided and used.

Nos. 82, 83 and 84 concern the generation of injurious dust or fumes and the handling of poisonous compounds. Where during grinding, cleaning, spraying or manipulation of any materials there is given off dust or fumes of such a character and amount as is likely to be injurious to the health of the operative all reasonable measures must be taken,by securing adequate ventilation or by the provision and use of suitable respirators or otherwise, to prevent the breathing of such dust or fumes. Where any persons are employed in processes in which a lead compound or other poisonous substance is used, there shall be provided for their use adequate facilities for washing, such as water, nail brushes, soap and towels, one basin or bucket for each five persons. No attempt should be made to remove paint from the hands with white spirit, since it dries out the natural oil and can lead to dermatitis. Some sprayers use a barrier-cream on face and hands before tackling overhead work. No mention is made in the regulations of the provision of milk drinks or other stimulants for sprayers. Where dust is given off during the process of grinding or cleaning suitable goggles or effective screens shall be provided to protect the eyes of the person employed.

Regulations Nos. 80 and 81 deal with first aid, shelters, and accommodation for clothes during working hours. Where more than 10 persons are employed, suitable first-aid boxes, adequately equipped, distinctly marked 'FIRST AID' and in the charge of a responsible person capable of giving first-aid treatment must be available or in the immediate vicinity. Adequate and suitable accommodation for taking shelter during bad weather, for depositing in a dry place clothing not worn during working hours, resonable arrangements for drying wet clothes, sufficient tables, benches or seats for taking meals, with facilities for boiling water, washing hands, and a convenient point and clearly marked supply of drinking water, must be provided.

According to the Lead Paint Regulations 1927, obligations of the painter include that overalls must be worn during the whole of the working period and that they must be washed at least once a week. Overalls must not be worn at meal-times. Clothing taken off during working hours should be kept well away from all painting materials, especially lead paint. Each person should wash his hands before partaking of food or leaving the premises. If deemed necessary, any person liable to come in contact with lead paint must present himself for medical examination. When a paint contains 5 per cent or more of soluble lead the container must be labelled 'this paint contains lead'.

Notice at end of roof constructed of fragile materials.

When doing work which involves obnoxious dust or fumes, goggles or a mask or both should be worn.

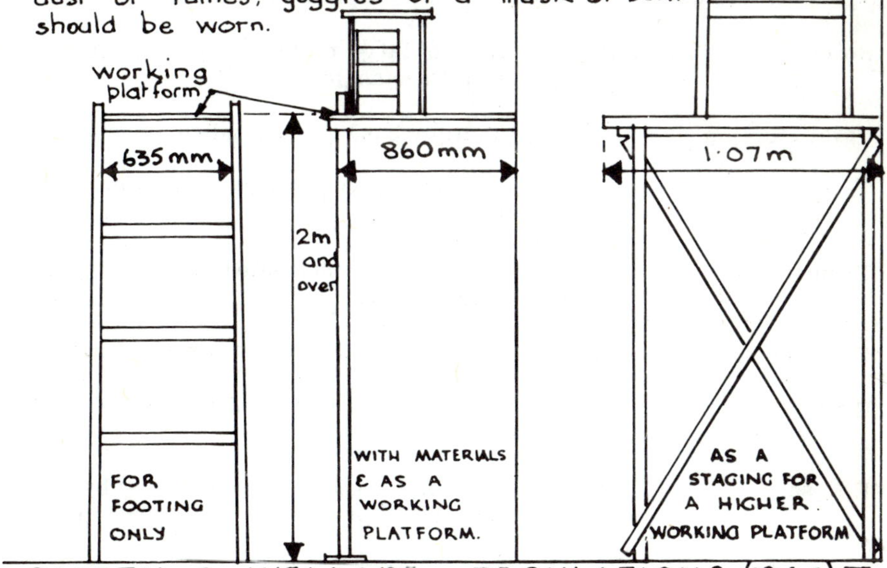

SAFETY & WELFARE REGULATIONS (1946) II.

Setting Out and Calculations

Painters' and decorators' work involves very little geometry and setting out, but it is nevertheless necessary to know the methods for solving a few simple problems as a help in the setting out of dados to staircases and ramps, centring a ceiling for a pattern paper or setting out for painted decoration. Vertical lines can be obtained by holding a plumb-line at any given point; horizontal lines by the use of a spirit level. A quick method of obtaining a perfect right-angle is to fold a large piece of paper, pierce a hole any distance from the fold unfold the paper and draw a straight line from hole to hole. The angle between fold and line will be a right-angle. Another method, generally used for setting out large sites, is to obtain a chalk line, measure off 3, 4 and 5 units, and set out the three lengths in the form of a triangle. Any triangle having sides in the proportion of 3, 4, 5 contains a right-angle. Another method, useful for ceiling work, is to draw at a given point a semi-circle of any radius. Using the same radius, mark off two points and from these points make two intersecting arcs. From the centre of the semi-circle, draw a line to the point of intersection. The larger the radius, the more accurate the line will be. The most satisfactory method of finding the centre of a rectangle is by drawing diagonal lines from corner to corner. For a ceiling with a projecting chimney breast two methods are suitable. The projection can be ignored and the lines taken from corner to corner or a line can be struck on the ceiling along the edge of the projection and the diagonal lines struck from the intersection of this line to the other two corners. To set out a dado on a staircase, inscribe arcs the height of the dado, using the intersections of the ramps as centres and striking lines tangentially from arc.

CALCULATIONS—AREAS

Figure	Formula
Square or rectangle	Length multiplied by breadth
Border around rectangle	Area of large rectangle minus area of small rectangle
Parallelogram	Multiply base by perpendicular height
Trapezium	Multiply average length of the side by height
Triangle	Half the base multiplied by the vertical height
Circle	π (= $3\frac{1}{7}$) multiplied by square of radius
Ring	Area of large circle minus area of small circle
Ellipse	Multiply long length (major axis) by main height (minor axis) by $\frac{4}{5}$—(Approximate method)

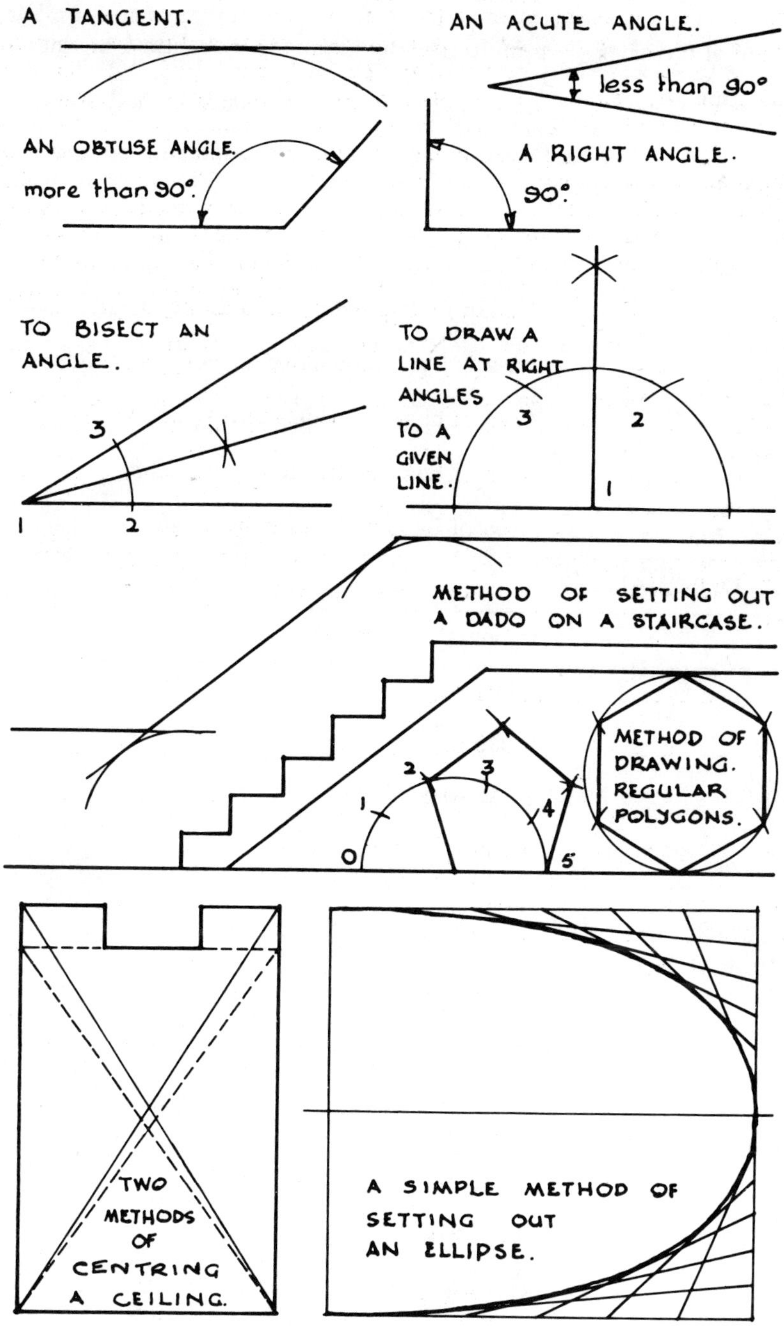
A TANGENT.
AN ACUTE ANGLE.
less than 90°
AN OBTUSE ANGLE.
more than 90°.
A RIGHT ANGLE.
90°.
TO BISECT AN ANGLE.
3
1
2
TO DRAW A LINE AT RIGHT ANGLES TO A GIVEN LINE.
3
2
1
METHOD OF SETTING OUT A DADO ON A STAIRCASE.
METHOD OF DRAWING. REGULAR POLYGONS.
0
1
2
3
4
5
TWO METHODS OF CENTRING A CEILING.
A SIMPLE METHOD OF SETTING OUT AN ELLIPSE.

SETTING OUT.

Regular Polygon	Find the area of one triangle and multiply by number of sides
Sphere	4π multiplied by square of radius
Curved Surface of Cylinder	π multiplied by diameter by vertical height
Curved Surface of Cone	π multiplied by radius of base by half slant height

CALCULATIONS - VOLUMES

Figure	Formula
Cube	Length multiplied by breadth by height
Prism	Area of base multiplied by vertical height
Cone	Area of base multiplied by $\frac{1}{3}$ height
Pyramid	Area of base multiplied by $\frac{1}{3}$ height
Cylinder	Area of base multiplied by height

UNITS—SI TO IMPERIAL

SI unit	Imperial equivalent
1 m	1·094 yd
1 mm	0·039 in.
1 litre	1·754 pint
1 kg	2·205 lb
1 g	0·035 oz

UNITS—IMPERIAL TO SI

Imperial unit	SI equivalent
1 inch	25·4 mm
1 foot	304·8 mm
1 yard	0·914 m
1 pint	0·57 litre
1 pound	453·4 g
1 ounce	28·34 g

Height in metres from skirting	Measurement in metres round walls including doors and windows																				
	10·0	11·0	12·0	13·0	14·0	15·0	16·0	17·0	18·0	19·0	20·0	21·0	22·0	23·0	24·0	25·0	26·0	27·0	28·0	29·0	30·0
2·0–2·25	5	5	6	6	7	7	8	8	9	9	9	10	10	11	11	12	12	13	13	13	14
2·25–2·50	5	6	6	7	7	8	8	9	9	10	10	11	11	12	12	13	13	14	14	15	15
2·50–2·75	6	7	7	8	8	9	9	10	10	11	11	12	13	13	14	14	15	15	16	16	17
2·75–3·0	6	7	8	8	9	9	10	10	11	12	12	13	14	14	15	15	16	17	17	18	18
3·0–3·25	7	8	8	9	10	10	11	11	12	13	13	14	15	15	16	17	17	18	19	19	20
3·25–3·50	7	8	9	10	10	11	11	12	13	14	14	15	16	17	17	18	19	19	20	21	21
3·50–3·75	8	9	9	10	11	12	12	13	14	15	15	16	17	18	18	19	20	21	21	22	23
3·75–4·0	8	9	10	10	11	12	13	14	15	16	16	17	18	19	20	20	21	22	23	24	24

Wallpaper measurement table: number of rolls required, normal patterned paper

Terms used in Paint Technology

ABRASIVE Material used either for rubbing down and smoothing surfaces or slightly roughening them to give 'key', e.g. glasspaper, emery cloth, steel wool, pumice stone and cutting pastes.

ABSORPTION Soaking up. Absorbent surfaces need priming paints with sufficient media to penetrate the interstices and also remain on the surface to form a foundation for the undercoating.

ADHESION Sticking or joining, by the use of an adhesive to penetrate holes in the two surfaces. This type of mechanical adhesion depends upon the strength of the key and is only satisfactory when the surfaces allow penetration by the adhesive. The first requirement of any adhesive is that it should wet the surface.

ALLIGATORING Deep and large fissures in paint which faintly resemble alligator hide. Movement of the substrate due to moisture may cause the paint film to crack, curl, and eventually fall off.

AMORPHOUS Substances so-termed have formless irregular shaped particles, e.g. china clay, some types of silica and vegetable black.

ANODE Term given to the positively charged body (or electrode) in connection with electrochemical attack of metals. The anode is always the part which is eaten away by such corrosion. Thus when mill scale is left on ironwork it becomes negative and the ironwork positive; the latter thus acts as an anode and rusts away.

BINDER (or Medium) Liquid part of paint capable of forming a solid film when spread thinly over a surface so holding the particles of pigment. Those used in paint-making are drying oils (linseed, tung, perilla and tobacco seed); modified semi-drying oils (castor, mineral, rosin and soya bean); and varieties of natural and synthetic varnishes. Frequently the binder consists of a mixture of oils and resins.

BLEACHING Loss of colour or fading caused by actinic rays of sunlight, alkali in marine atmospheres, or chalking. It is more marked in strong dark colours than in light ones.

BLEEDING Discolouration of a painted surface due to materials penetrating freshly applied paint. Coal tar dyes, bituminous materials, resin contained in knots, and some products of combustion, such as soot, behave in this way. If they cannot be removed they should be sealed with spirit varnish or aluminium wood primer.

BLISTERING On a painted surface, caused by moisture underneath the paint film expanding through a sudden rise in temperature. Moisture may be present in the fabric of a building and unless it can evaporate it will move around and find the points of least adhesion of the paint film.

BLOOMING Cloudy or milky appearance of varnish, enamel and gloss paints due to the settling of moisture on the surface. Two types occur: (1) 'pinhole' or 'crater' in the form of pinholes or small craters on the surface; (2) 'crystalline' caused by the deposit of ammonium sulphate crystals on the finish. The latter type of blooming can usually be removed by sponging with warm water and vigorous rubbing.

CATALYST Substance which brings about a chemical change in a material whilst itself remaining unchanged. Examples are the 'two pack' materials in which the paint hardener (catalyst) is added to the base immediately before use.

CATHODE Negatively charged body or electrode, in electrochemical attack of metals.

CHALKING Powdery condition of a painted surface indicating that the binder has partially or completely broken down. Fading or bleaching of the colour is usually the first indication. Some paint products exploit this type of weathering as the paint looks cleaner longer, and for repainting rubbing down the surface is the only preparation required. Severe chalking is called EROSION.

CHECKING See Cracking

CISSING Paint gathering in some places and receding from others during application. Caused by surface tension and met with when applying graining colour in water or a varnish over an oily ground.

CLOUDING Defect of varnish work due to badly made varnish.

COLLOIDAL Substances so-termed do not form solutions or suspensions but exist in a state of such minute particles that when dispersed in a liquid they do not settle out, e.g. glues in water and resins in varnishes.

COLOUR Sensation produced by visible light upon the eye. When red, orange, yellow, green, blue and purple coloured lights are mixed together white light is produced, and any two coloured lights will together produce a brighter coloured light than the originals. When red, orange, yellow, green, blue and purple pigments are mixed, black is produced, and when two pigments are mixed the resultant colour is less intense than the original ones.

CONCENTRIC CRACK FORMATIONS Defect brought about by a combination of incompatability of the paint films and sudden extremes of temperature.

CONDUCTIVITY The rate at which heat will pass along or through a specific material.

CONSISTENCY Degree of thickness of a paint i.e. round, stout or stiff.

CONTINUOUS PHASE The larger proportion of a liquid medium in an emulsion.

CORROSION Chemical attack which all metals undergo when exposed to the action of oxygen, water, salts, acids and alkalis. With iron and steel, it is referred to as rusting. Pre-treatment before painting is most essential.

CRACKING Defect in a painted surface generally due to incompatability of the various layers of paint, e.g. application of a sharp or quick drying paint over a slower drying or more elastic one; or movement of the surface or substrate itself. The formation of the cracks usually indicates the cause and degree of the defect. When the surface is covered with fine cracks which do not penetrate the paint or polish film it is known as CHECKING.

CURTAINING or SAGGING Paint defect caused by uneven, or in most cases too generous, application. Shows as thick uneven ridges suggestive of a draped curtain.

CRYSTALS Regular shaped particles, e.g. graphite, asbestine, antimony white.

DELIQUESCENCE Change undergone by certain substances which absorb so much water that a saturated solution is formed. Calcium chloride and sodium hydroxide (caustic soda) behave in this way.

DISPERSE PHASE The smaller proportion of a liquid medium in an emulsion.

DRIER Solid or liquid agent in paint with ability to change it from a fluid to a hard film. Evaporation of the thinner and partial oxidation of the medium produces a state of 'touch dryness'. Complete oxidation and polymerization produces the 'hard dry' state. With some materials the oxidation process is not required.

ADHESION.

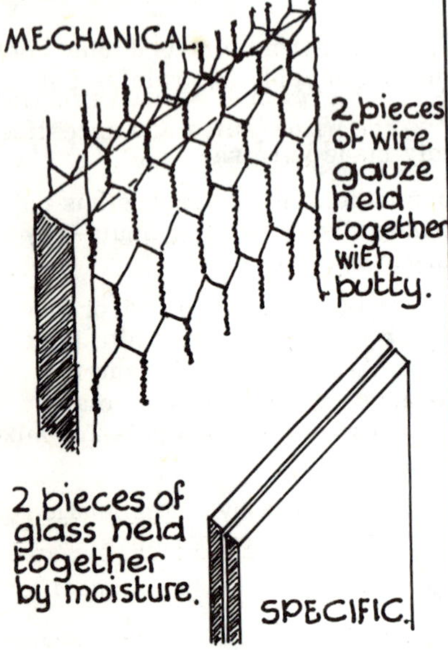

CORROSION.

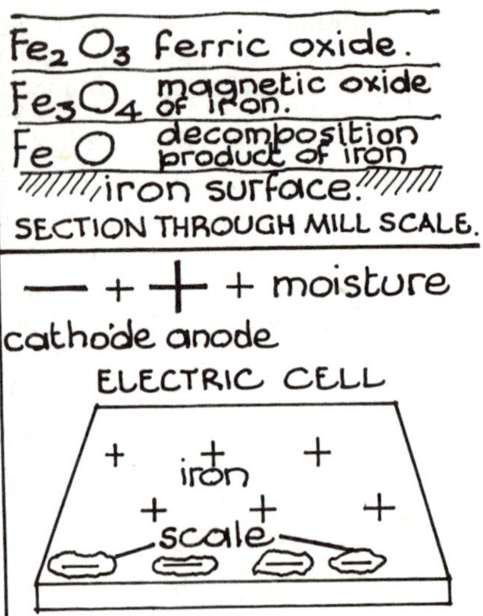

RADIATION & CONVECTION.

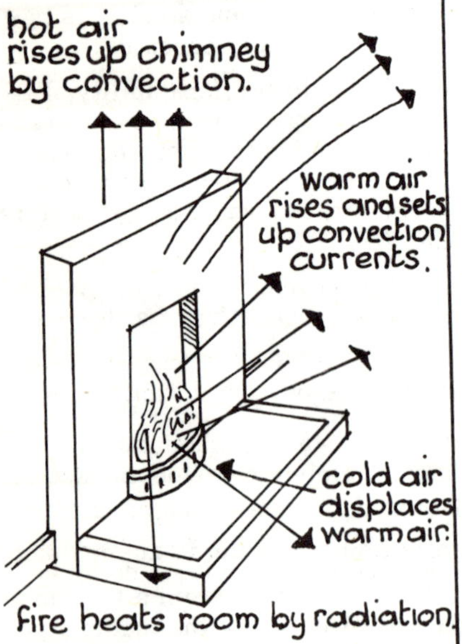

TESTING PAINT.

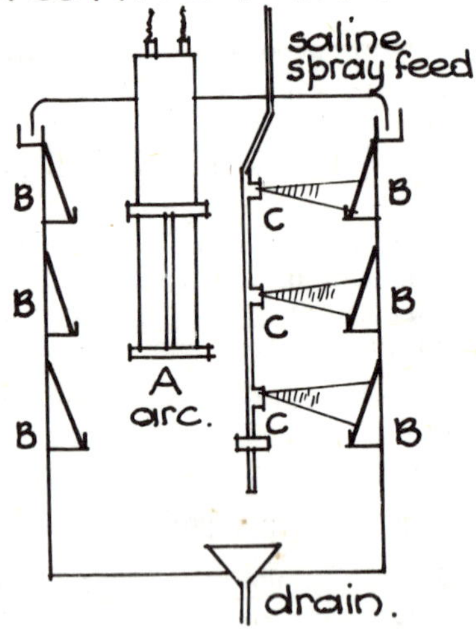

CROSS SECTION THROUGH WEATHERING APPARATUS.
A - arc. B- panels under test. C - spray.

EFFLORESCENCE White fluffy crystalline deposit on the surface of plaster, cement and bricks, caused by the drying out of salts.

ELASTICITY Ability of a paint film to respond to movements of the substrate without cracking, or to the stresses which are set up during drying.

EMULSION Fluid mixture consisting of fats, waxes or oils and water stabilized with an agent such as soap which reduces the surface tension between the two fluids. Emulsion paints are either of the oil in water or water in oil type.

EROSION see Chalking

EXTENDERS White minerals added to paints which give bulk without affecting the colour. They are selected to give 'tooth' to the paint (silica and asbestine) or as suspending agents for heavier pigments (silica and china clay).

FEEDING Thickening of a paint (sometimes to the extent of making it unusable) due to the reaction of certain basic pigments such as zinc oxide with the acid in the binder.

FLAKING Paint or distemper may flake due to moisture, or lack of key and tooth.

FLASHING or SHUCKS Shiny patches of paint on a flat finish due to over brushing during application.

FLASHPOINT Temperature to which an inflammable liquid must be heated before its vapour ignites when in contact with a flame—a physical means of testing its purity.

FLOATING Separation of pigments in a paint either in the can or the dried film. It is caused by 'bad wetting' of the pigment during manufacture and can be aggravated by overthinning.

FLOCCULATES Agglomerates or clusters of pigment particles due to incomplete wetting of the pigment.

FLUID A material in a mobile state; its particles move freely over or with each other enabling it to accommodate itself in any container. Thinners reduce the consistency of paints and make them more fluid, i.e. easier to apply.

HARDNESS Ability of a paint system to withstand hard knocks, usage or mechanical abrasion. Stringent tests are made to check the resistance of paints e.g. the scratch test and special apparatus such as the hardness rocker, the guide tube impact and the lever arm impact apparatus is used.

HYDROMETER Instrument for measuring the specific gravity of liquids e.g. water 1.0, white spirit 0·789 and turpentine 0·864.

HYGROMETER Instrument for measuring moisture content of surfaces such as plaster and wood.

HYGROSCOPIC Absorbing moisture readily, e.g. substances such as caustic soda and calcium chloride. Incorporated in the fabric of a building such substances cause surfaces to become permanently damp or to appear damp in humid weather.

LATTICE A surface built up of strips crossed diagonally, or horizontally and vertically, such as that formed when an emulsion dries, in which the material bridges the gaps through which the moisture disperses.

LEACHING To saturate, with the object of melting or dissolving and thus removing salts or other deleterious matter from a surface.

LIVERING Condition in which the thick leathery condition of the paint renders it useless. It may be due to the reaction between the pigment and the binder as in 'feeding' or the formation of polymers during storage; and if the mass at the bottom of the container is very hard it is caused by 'bad wetting' of the pigment.

LOSS OF FACE Condition which occurs when a painted surface loses its gloss due to a rapid fall of temperature, or is applied over a surface whose suction has not been completely satisfied.

MOISTURE MOVEMENT Moisture in a building moves about, largely influenced by variations in temperature. Moisture movement upwards and sideways is capillarity, downwards is soakage or seepage, and outwards evaporation.

OPACITY Ability of a material to hide or obliterate the surface over which it is applied. It should not be confused with covering power which merely refers to the area over which a material can be spread.

ORGANIC SOLVENTS Volatile spirits which dissolve resins and waxes enabling them to be mixed with each other or with oils. Turpentine is used for waxes and resins, pine oil for resins, alcohols for cellulose materials and xylene for bitumens. The dissolved part of the solution is known as the solute.

OXYNS Combination of molecules of oxygen and linseed or tung oil formed during the drying of an oil paint.

PAINTING DEFECTS Changes which occur in the condition of paint films, either immediately after application or over a period of time, fall generally into four main categories: Change of colour (Bleaching, Bleeding); Loss of gloss (Blooming, Chalking, Clouding, Loss of Face); Non-drying (Tackiness or Sweating, Wrinkling, Saponification); Surface defects (Blistering, Cissing, Cracking, Alligatoring, Curtaining or Sagging, Runs or Tears, Flaking, Flashing or Shucks,

POWDERING (Due to Efflorescence) Powdering and eventual disruption of the paint film may be caused by salt deposits developing on the surface underneath the paint film. Removal of all loose particles and salts and an interim period for further drying out are essential.

pH VALUE indicates the acidity or alkalinity of a solution (which depends on the concentration of hydrogen ions present). Low pH values (0-6) show acid solutions; 7 denotes neutral; and high pH values (8-14) alkaline reactions.

PIGMENT Solid particles in paint which give opacity and colour and in some cases protection against chemical attack. The most important are of inorganic origin, being derived from iron, lead, zinc, chromium, titanium, copper, etc.

POLYMERIZATION Complex chemical change which occurs in paint and varnish films when they change from touch dry to hard dry. Normally oil paints dry in three stages: (1) evaporation of the thinner, (2) oxidation of oil to form oxyns, (3) polymerization. With synthetic materials the oxidation stage is often not necessary.

REFLECTION The process of throwing back light, heat or sound from a surface.

REFRACTION When a ray of light or heat is bent or deflected from a direct course.

RUNS or TEARS Isolated places where too much paint has accumulated e.g. at the intersections of mouldings.

SALT Substance in which the hydrogen atoms of an acid have been replaced wholly or partly by a metal or a metallic radical.

SEDIMENTATION Method of separating a pigment into particles of different sizes. Also refers to the settling of pigments in paint during storage.

SOLUTION Liquid in which a solid has been dissolved.

SPECIFIC GRAVITY Relation of the density (weight per unit volume) of any material to the density of water which is 1 kg per litre; a physical method of checking the identity of materials such as pigments.

SPECTRUM Image consisting of bands of different colours produced by passing white light through a prism. The spectrum of sunlight shows the colours red, orange, yellow, green, blue, indigo and violet.

SPECTRAL ANALYSIS Method of identifying substances by the wavelengths they emit.

SPREADING CAPACITY Area over which a painting material can be spread. Depends upon the porosity of the surface, thickness of the material, temperature and skill of the operative. Spreading capacity is quoted either in square metres per litre or, for paste materials, square metres per 50 kg.

SURFACE TENSION Property of a liquid, causing it to act like an elastic skin at the surface. The rise of liquids in fine tubes is due to this and is called capillarity.

SUSPENSION Very fine particles of solid material floating or suspended in a liquid. All paints are suspensions.

SWEATING See Tackiness

SYNTHETIC RESINS Manufactured resins made from nonresinous substances. Types which give paint media special properties are phenolic, alkyd, acrylic styrene, vinyl, melamine, maleic, epoxy and urea.

SYNTHESIS Combining of parts to make a whole, or the process of building up a substance (such as a resin) by combining the elements of which it is formed.

TACKINESS or SWEATING Permanently sticky finish (in varying degrees) due to oily matter exhuding through the finishing coat of paint or varnish. Caused by applying the finishing coat too soon upon an undercoat which has not properly dried. WRINKLING often occurs through this fault. Sometimes paints do not dry because of the reaction between the alkali from the building and the fatty acids of the paint. Usually blisters, discolouration and dark tears down the surface indicate this action which is known as esterification.

TEARS See Runs

TESTING OF MATERIALS Besides the checking of materials for purity all paints are subjected to tests for adhesion, elasticity, gloss, hardness, fineness of grinding and flow under strictly controlled conditions. Apparatus used for these tests includes the sheen gauge for fineness of grinding; the hardness rocker for hardness and drying, and there are bend, scratch and impact tests for adhesion.

TEXTURE Inherent quality of a surface, e.g. roughness or smoothness, brought about by the particular arrangement of its tissues.

THINNERS Liquid used to reduce the viscosity of a painting material. As a general rule turpentine, white spirit, and solvent naphtha are used for oil paints, amyl and butyl acetate for cellulose materials, methylated spirit for shellac and polishes and water for distemper and emulsions.

TOOTH Roughness or coarseness in a dried paint film which assists in the application and bonding of a subsequent coat of paint.

TOXIC Poisonous. (Waterproofers used in a confined space give off toxic fumes).

VISCOMETER Instrument for measuring the viscosity of a liquid. Types in use are U tube, efflux, torsion falling sphere and cup and ball.

VISCOSITY Consistency (thickness) or resistance to flow of a liquid. The unit of dynamic viscosity is the newton second per square metre (Nsm^{-2})

VOLATILE SPIRITS Liquids used for thinning paint so reducing the mixture to a workable consistency. They evaporate quickly and do not form any part in the paint film.

WARNING OR SAFETY COLOURS These are the colours recommended for accident prevention and to meet certain emergencies that may arise in factories and other industrial buildings. Signal Red means stop, fire fighting equipment, explosives or other dangerous substances. Orange Yellow or Golden Yellow means warning of danger or hazards. Mid-green indicates safety equipment, routes to be used in an emergency or areas of safety. To give added contrast to these colours the Red should be chequered with white—except for fire fighting equipment and stop buttons which should be shown with a white surround. The orange yellow should have a black contrast shown by the use of alternate diagonal stripes; and the mid-green contrasted with white, white lettering on a green ground or symbols in green on a white ground.

WARNING OR SAFETY SYMBOLS Coupled with the safety colours certain geometric shapes are used for specific purposes:- CIRCLES prohibit or command, TRIANGLES warn of hazards or danger, RECTANGLES give information.

WATER OF CRYSTALLIZATION Water which is present in hydrated crystals. Especially important in the manufacture and setting of plasters.

WEATHERMETER Instrument for measuring the resistance of paints to simulated weather conditions.

WRINKLING See Tackiness

Index